MUSCLES, MUSCLES, MUSCLES

Muscle Anatomy and Physiology for Aspiring Young Doctors

CONTENTS

MUSCLE BASICS

Muscles are a fascinating and critical part of our body. In this section, we will learn what muscles are, what they are made of, how they work, and how they are attached to the body.

WHAT IS A MUSCLE?

Your heart, lungs and stomach are called organs. The human body has many organs.

However, a muscle is not an organ. A muscle is a tissue in your body that can contract and relax to produce movement. There are four different types of **tissues** in the body and one of them is the muscle.

Muscles are essential for almost every action you perform, from walking and lifting to smiling and talking.

Even superheroes must rely on their muscles to perform all the heroic and amazing things they do.

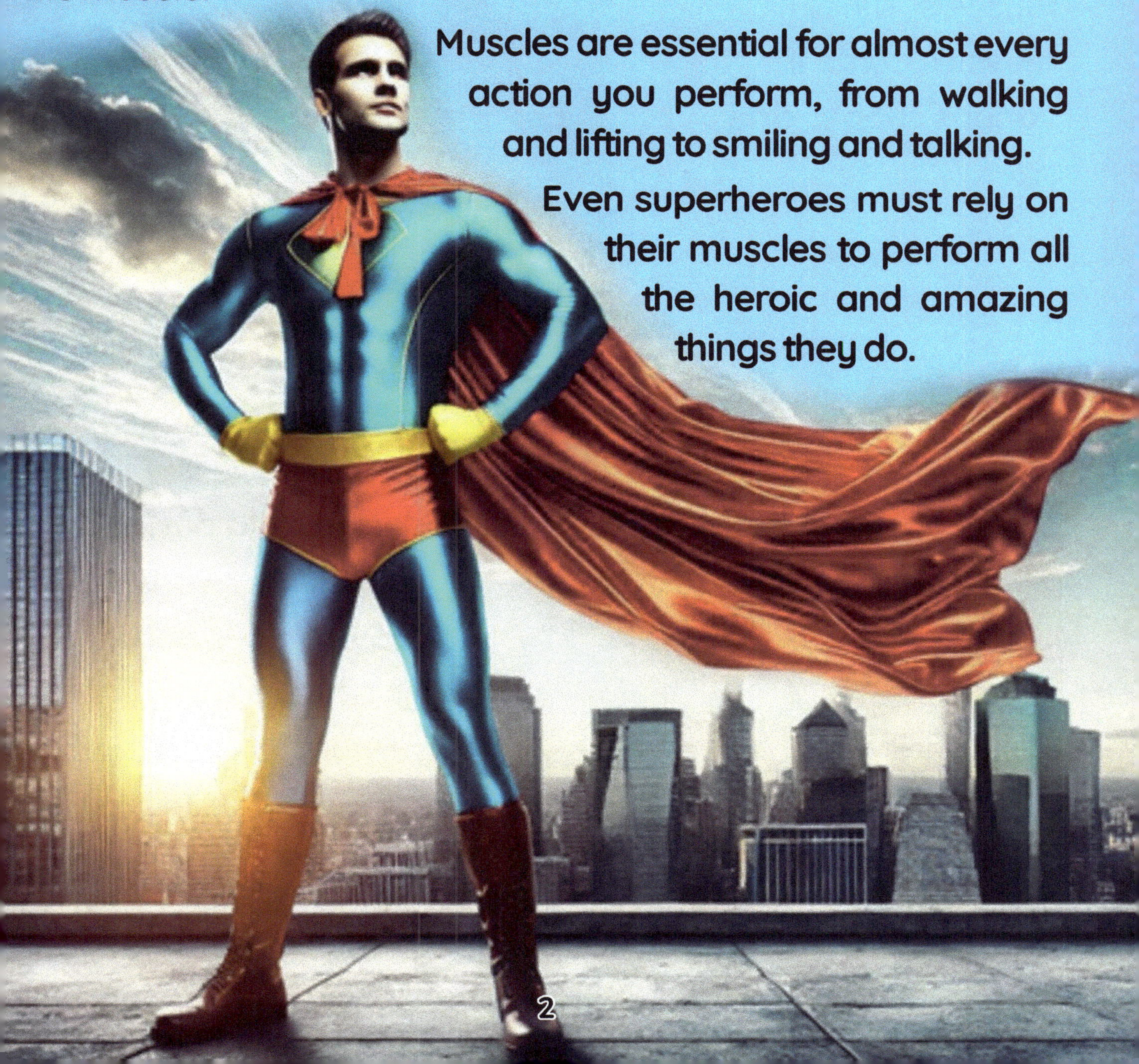

HOW MANY MUSCLES ARE THERE?

The human body has more than **600** muscles! These muscles come in various shapes and sizes.

Each muscle is designed to perform specific functions, from moving your bones and maintaining posture to circulating blood and aiding in digestion.

THE IMPORTANCE OF MUSCLES

Muscles have many important roles and functions. Let's review some of them.

1. Movement

Muscles are responsible for all voluntary movements, such as walking, running, lifting, and typing. They contract and relax to move the bones and joints.

2. Stability and Posture

Muscles help maintain posture by keeping the body upright and balanced. They support the spine and other structures, preventing falls and injuries.

3. Joint Protection

Muscles provide stability to joints, reducing the risk of dislocations and other injuries. They help absorb shocks and distribute loads evenly across joints.

4. Heat Production

Muscle activity generates heat, which is crucial for maintaining the body's temperature. Shivering, for example, is a muscle activity that produces heat.

HOW MUSCLES WORK

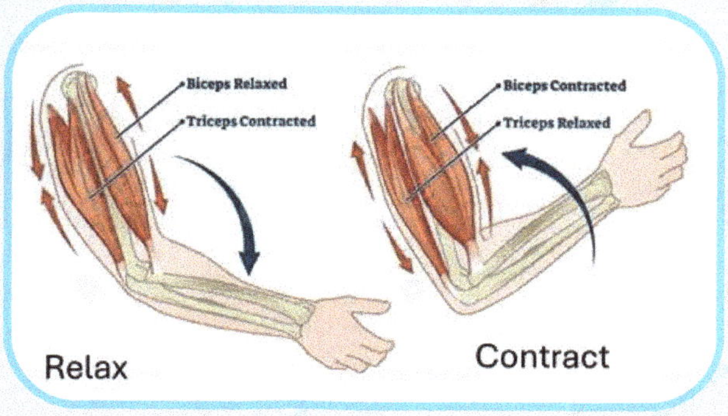

Muscles work by contracting (shortening) and relaxing (lengthening). When a muscle contracts, it pulls on the bones or tissues it's attached to, creating movement. This process is controlled by signals from your nervous system.

An example of this is when you lift something heavy your muscles tighten or contract. When you place it down, your muscle relaxes.

WHAT ARE MUSCLES MADE FROM?

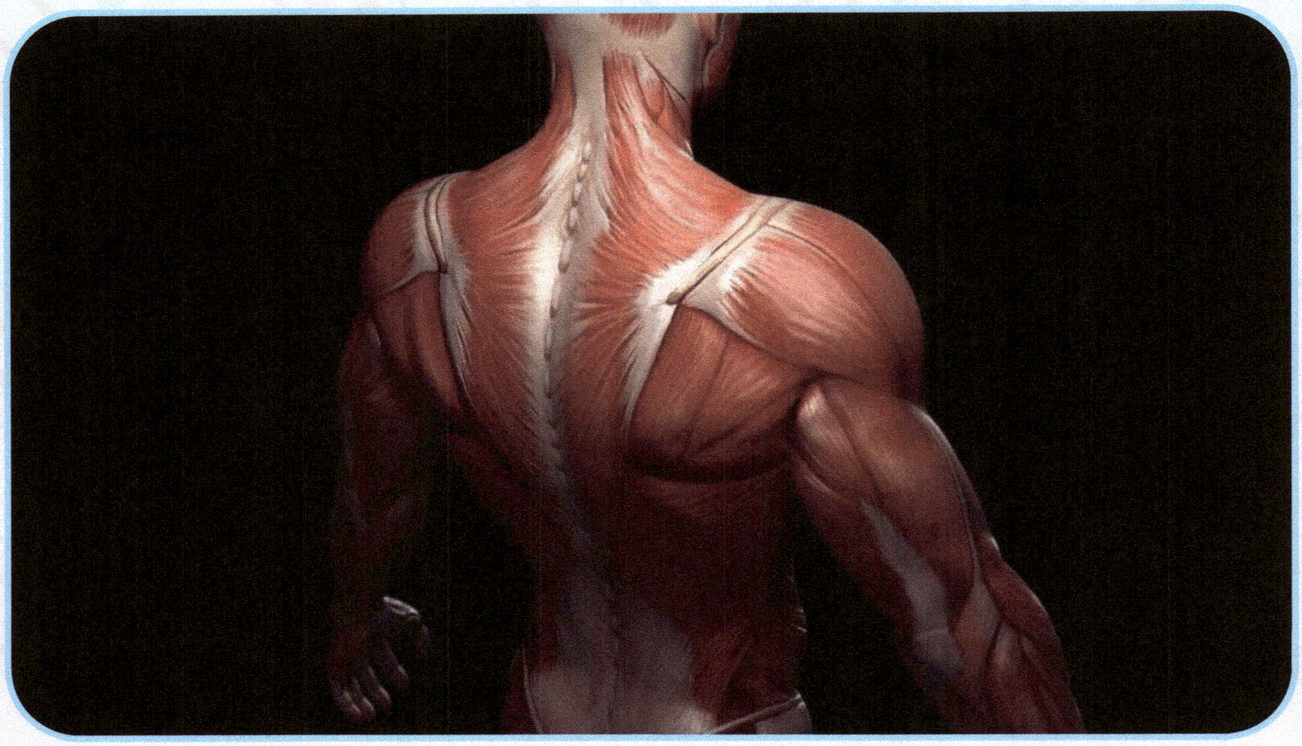

Muscle fibers are the main component of muscle tissue. Muscle fibers are long, cylindrical cells that can contract to produce force. Muscle fibers are made of protein. Many protein fibers come together to build a single muscle.

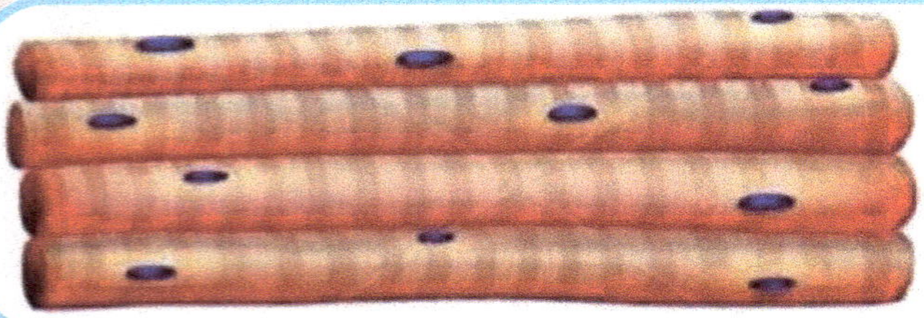

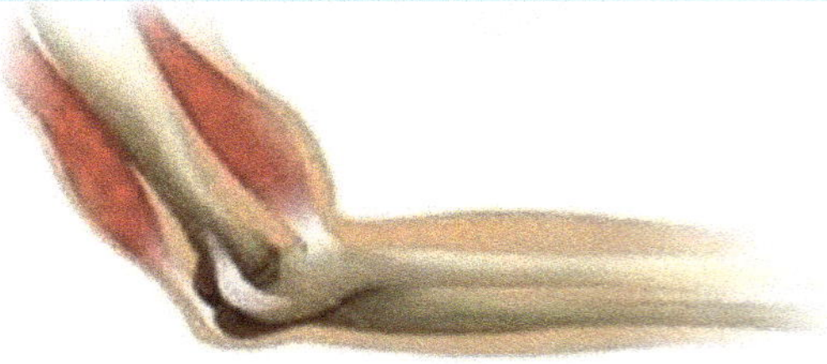

COMPARING CABLES AND MUSCLES

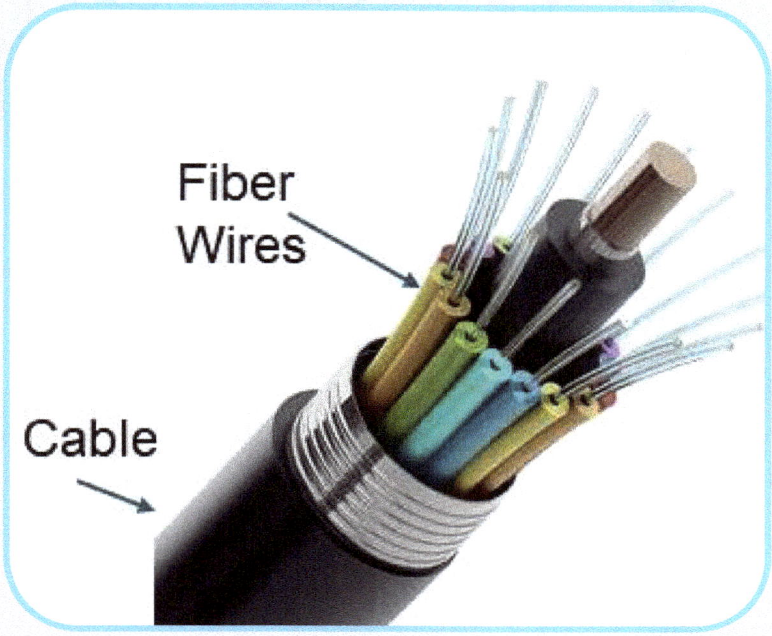

Cables are all around us. You can find them connected to your computers, video games, lamps, televisions, and more. Inside a cable are a bunch of wires. Some of these wires are called fiber optics. A single cable can be made up of a bundle of fiber optic wires.

CABLES AND MUSCLES

Muscles are like cables. But muscles don't have wires inside them, they have muscle fibers. Many fibers made from proteins come together to form a muscle bundle.

Many muscle bundles come together to create a single muscle.

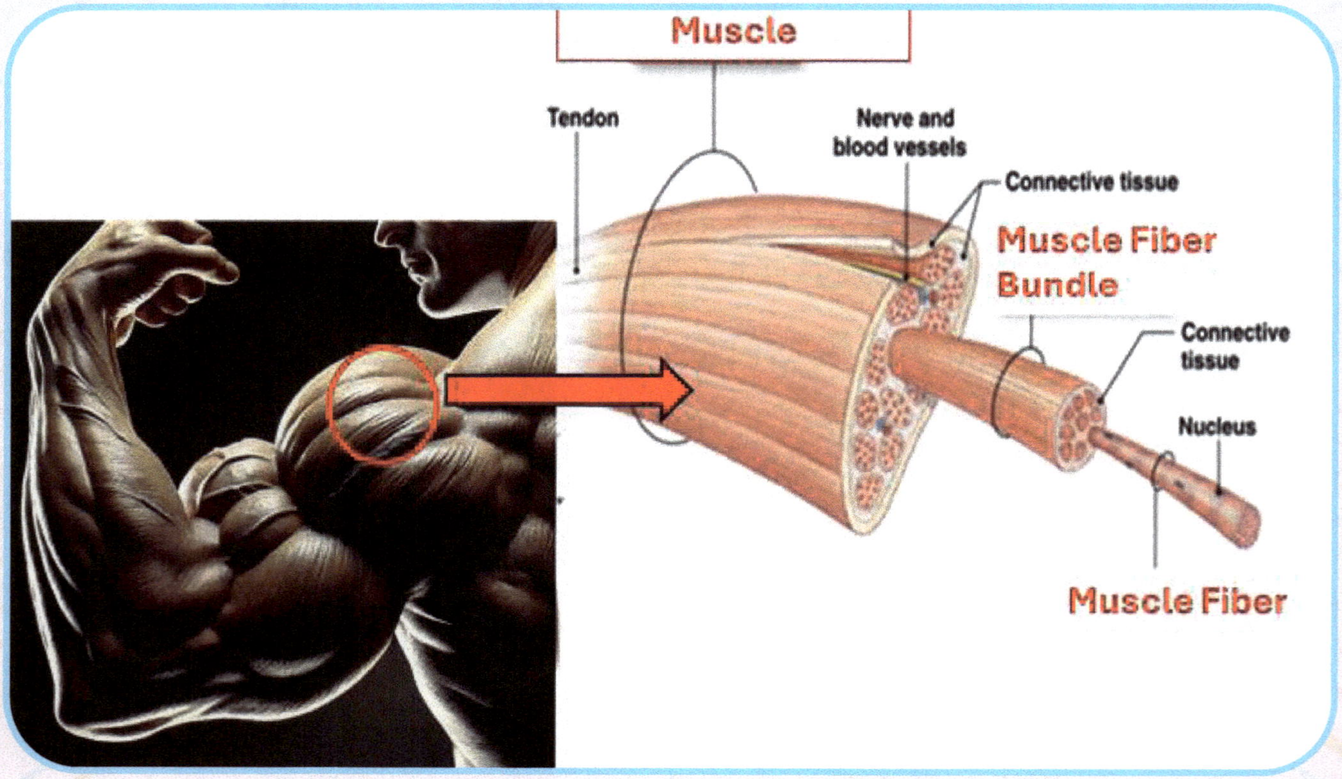

HOW MUSCLES GET ACTIVATED

There are many activities that happen very quickly each time we use our muscles. It all begins with a single thought.

1. An electrical signal is sent from the brain.

2. The electrical signal travels to the muscle fibers.

3. A chemical reaction takes place between the nerves and the muscles.

4. Sodium flows in the muscle and potassium flows out.

5. The muscle moves and contract.

6. The muscle relaxes.

HOW ARE MUSCLES ATTACHED?

Muscle Fibers: Muscles are made up of many fibers. These fibers come together to form a bundle.

Tendons: At the ends of muscles, the fibers converge into tendons, which are strong, fibrous tissues. They have a white-colored look. Tendons connect muscles to bones, allowing for the transfer of force when the muscle contracts.

Bone Attachment: The tendons are anchored to the bones at specific points known as entheses. These attachment sites are usually at the ends of long bones, near joints.

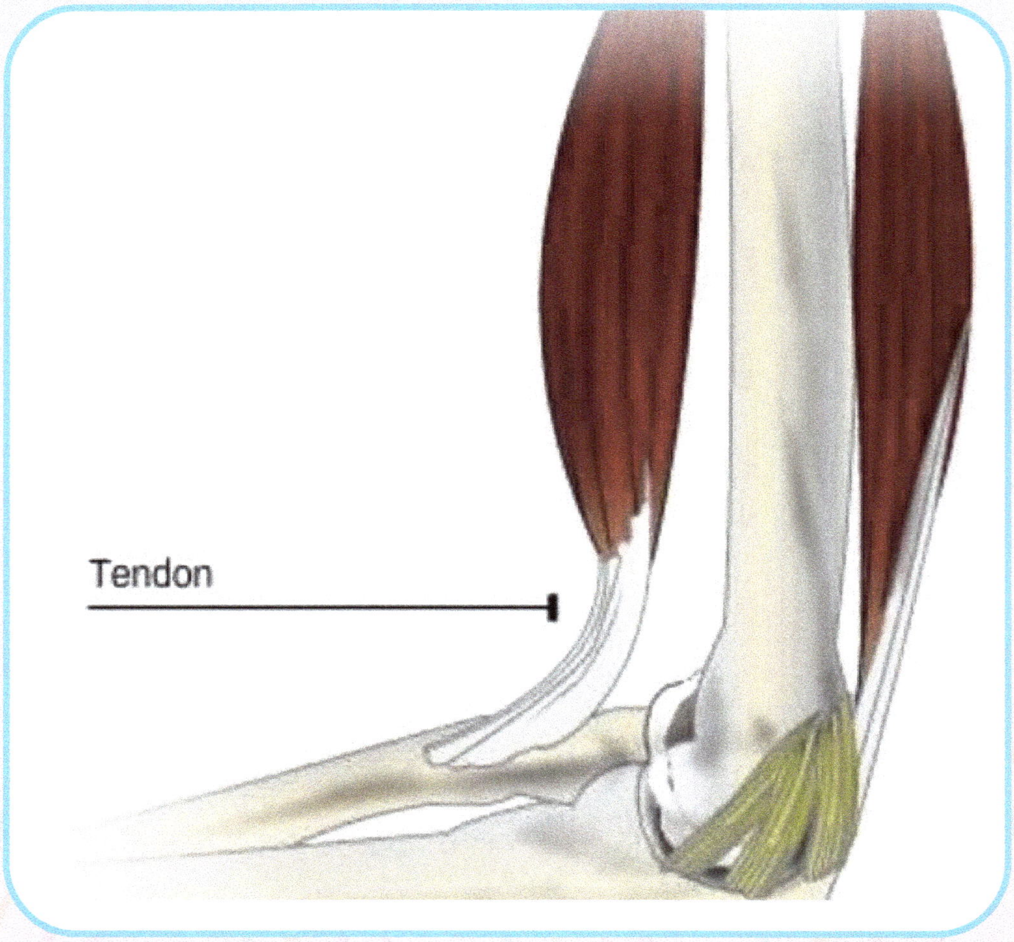

TYPES OF MUSCLES

Although there are hundreds of muscles in the human body, all of them can be categorized into just three major types. This section will review these categories. Let's find out what they are.

SKELETAL MUSCLES

I. **Skeletal Muscles:** These muscles are attached to bones and are responsible for voluntary movements. They help you walk, run, lift, and perform other activities.

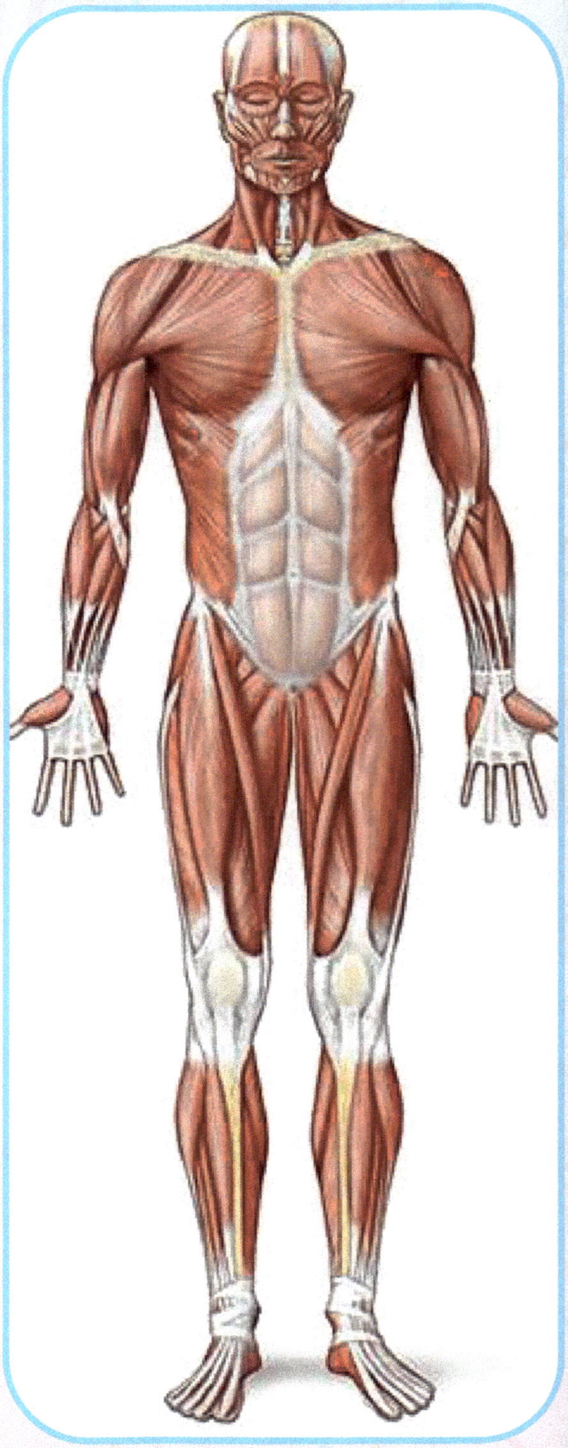

Detailed View

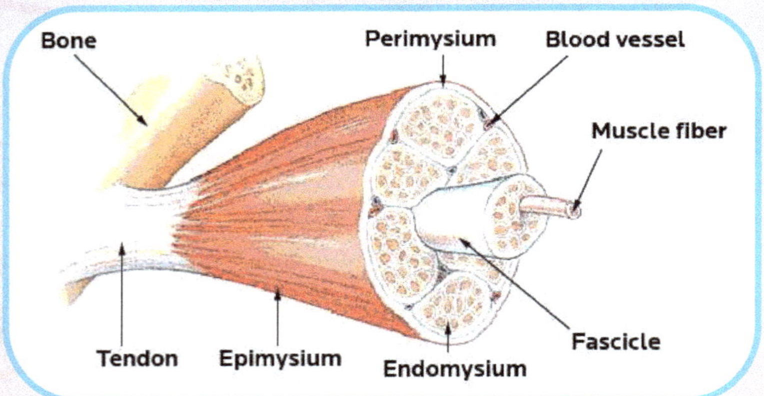

Definition

"A voluntary muscle is a muscle that you can control with thought."

SMOOTH MUSCLES

2. **Smooth Muscles:** Found in the walls of internal organs like the stomach, intestines, and blood vessels. These muscles control involuntary movements such as digestion and blood flow.

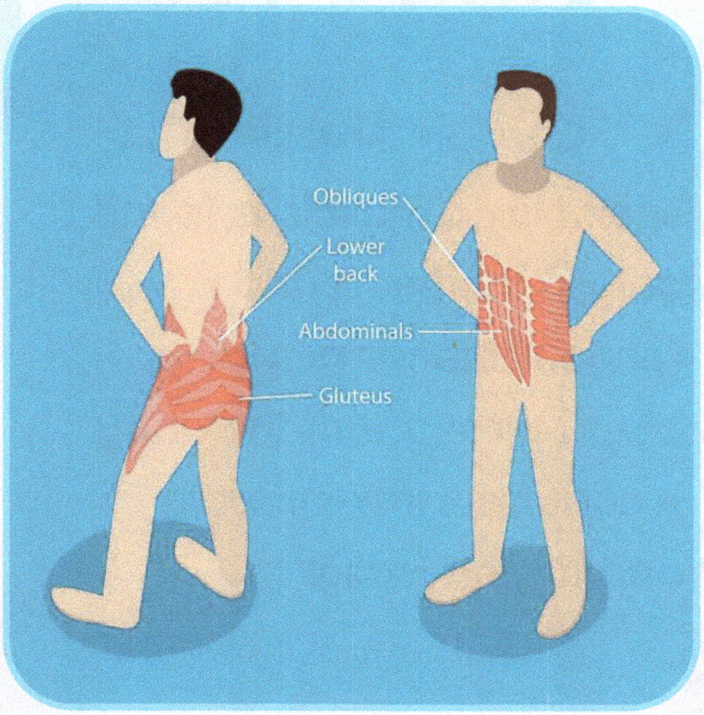

Detailed View

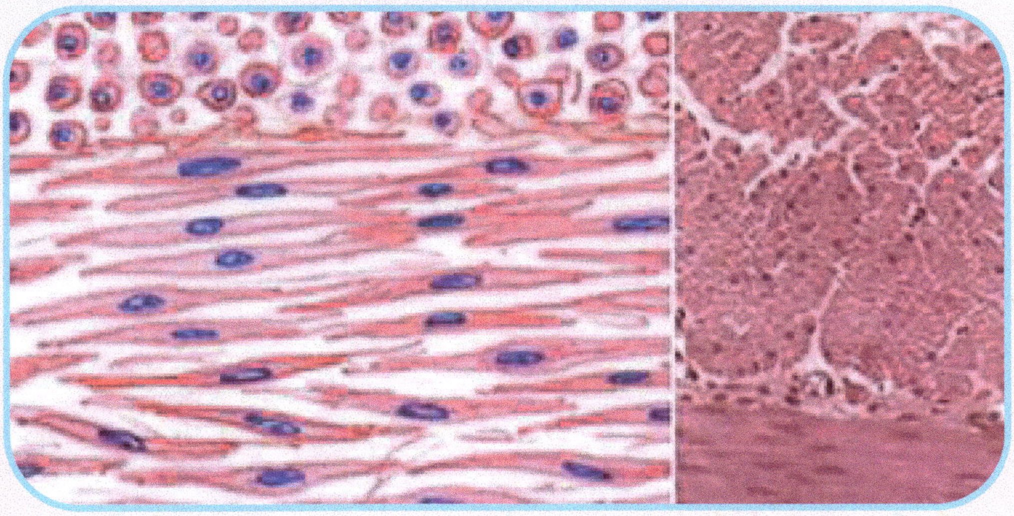

Definition

"An involuntary muscle is one that you cannot control with thought."

CARDIAC MUSCLES

3. Cardiac Muscle: This specialized muscle is found only in the heart and is responsible for pumping blood throughout the body. It contracts involuntarily and rhythmically.

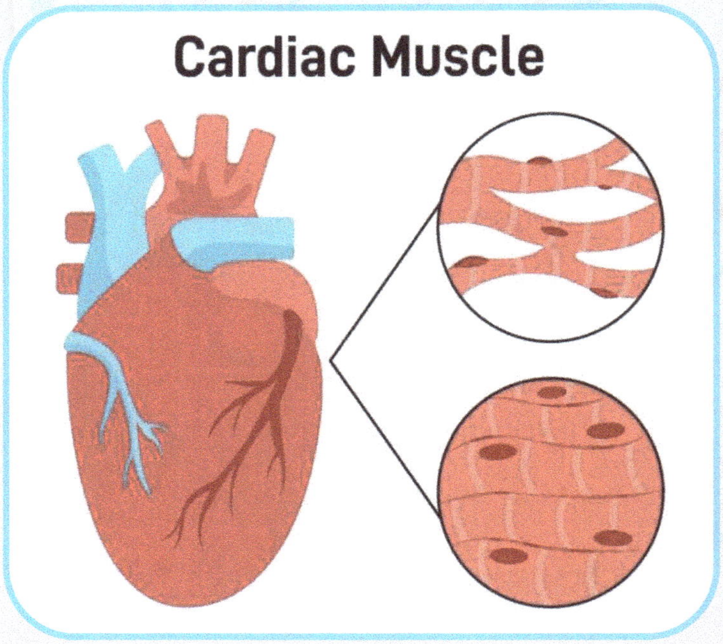

Detailed View

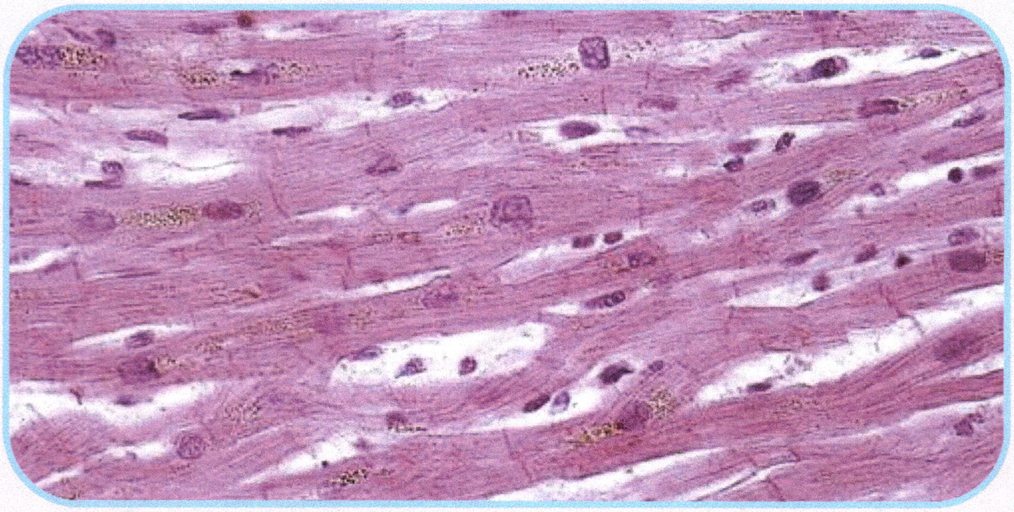

MUSCLE NAMES

Everything has a name and with over 600 muscles in the body it is important to name each one. This section is all about naming our muscles.

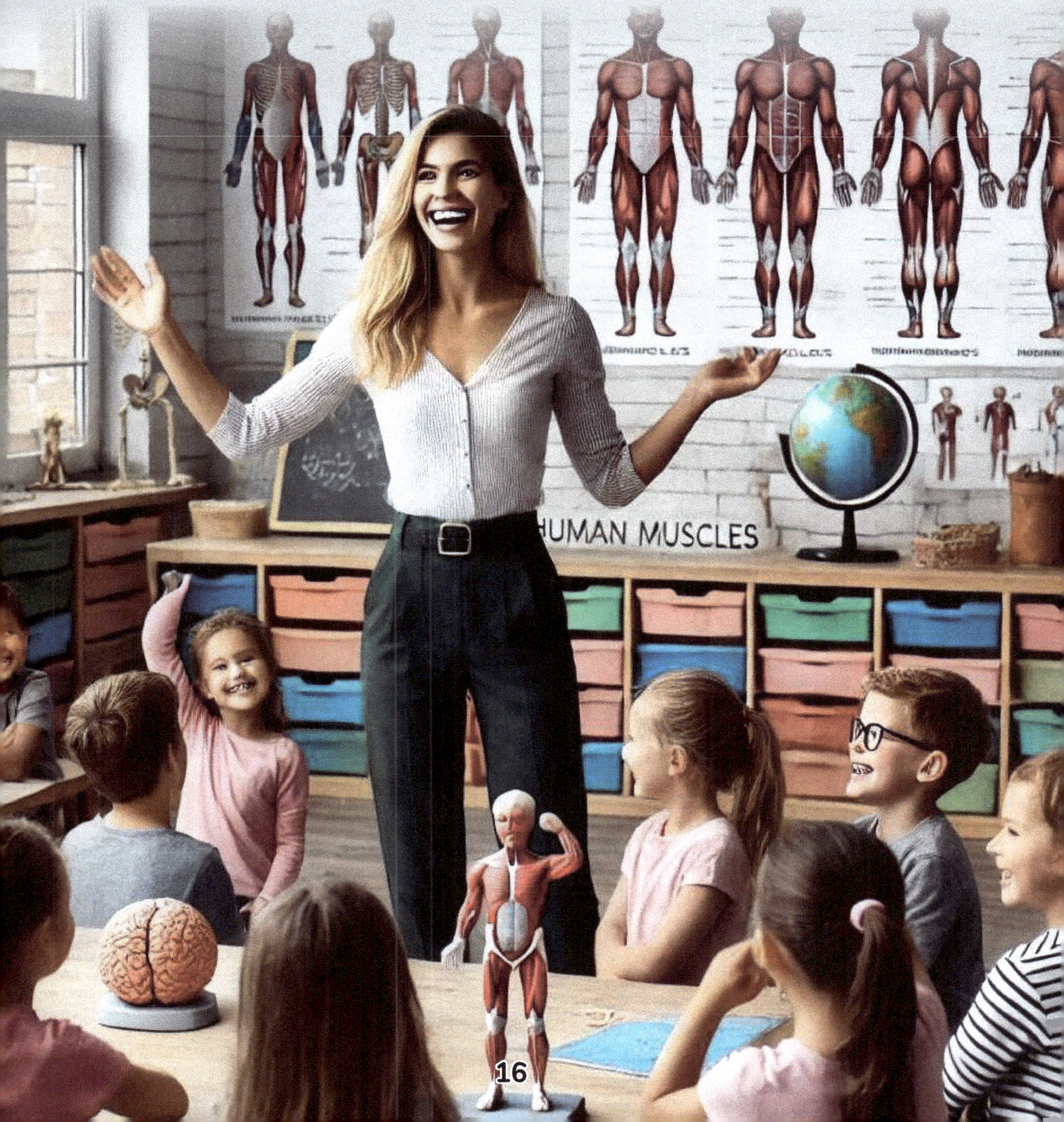

THE BASIC NAME OF MUSCLES

The names of muscles were developed hundreds of years ago by early anatomists, and we continue to use them today. They help to provide a clear and standardized way to identify and describe muscles.

These names were derived from the Latin and Greek language and may sound usual or even difficult to pronounce.

So, let's first start with the more common and easier names that most people know.

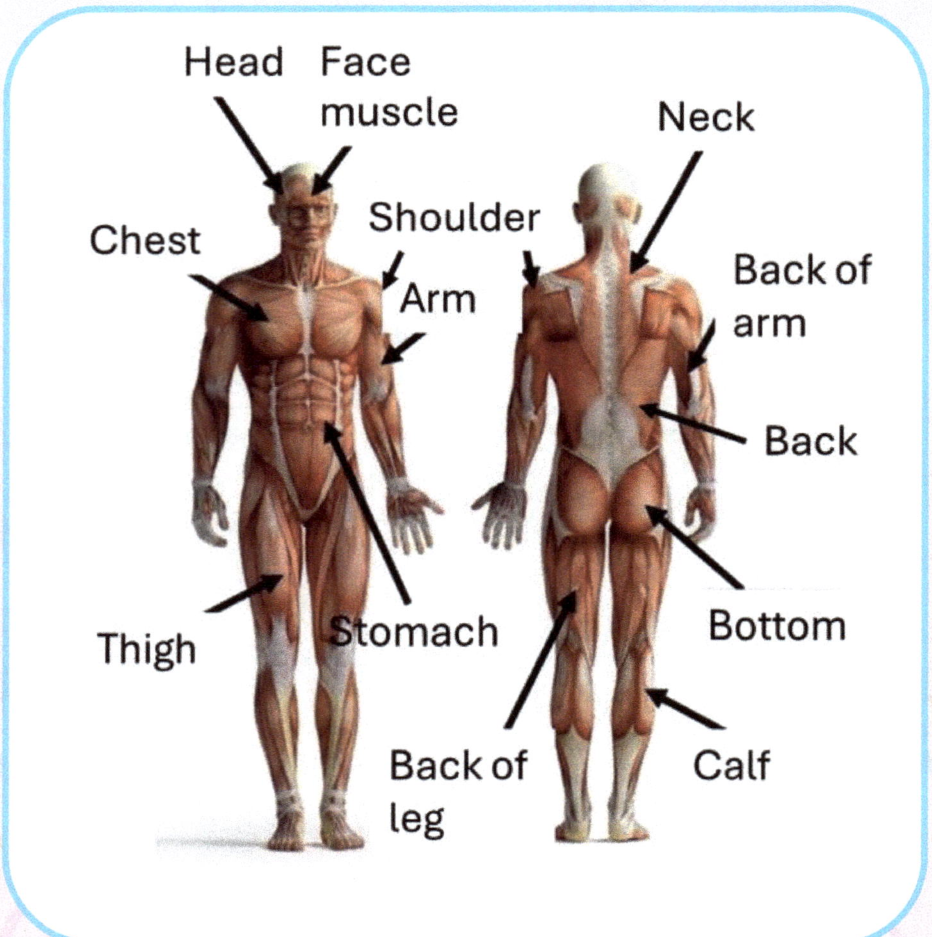

MODERN NAMES OF MUSCLES

The next step is to learn the more modern names used for certain muscles today.

Biceps – These two muscles make up the upper arm muscles.

Abs – The stomach muscles are called abdominals or "abs" for short.

Quadriceps – This is the name of the thigh muscles, called "quads" for short.

Traps – These are part of the lower neck muscles.

Forearm – this is the name for the lower portion of the arms.

Triceps – The back of the arm comprises three muscles called triceps.

Upper and Lower Back – The back muscles can be divided into upper and lower back muscles.

Glutes – The two bottom muscles are called glutes.

Hamstrings – The upper back leg muscles are called hamstrings or "hams."

Pectoralis – The chest muscles or "pecks" for short.

MODERN NAMES OF MUSCLES

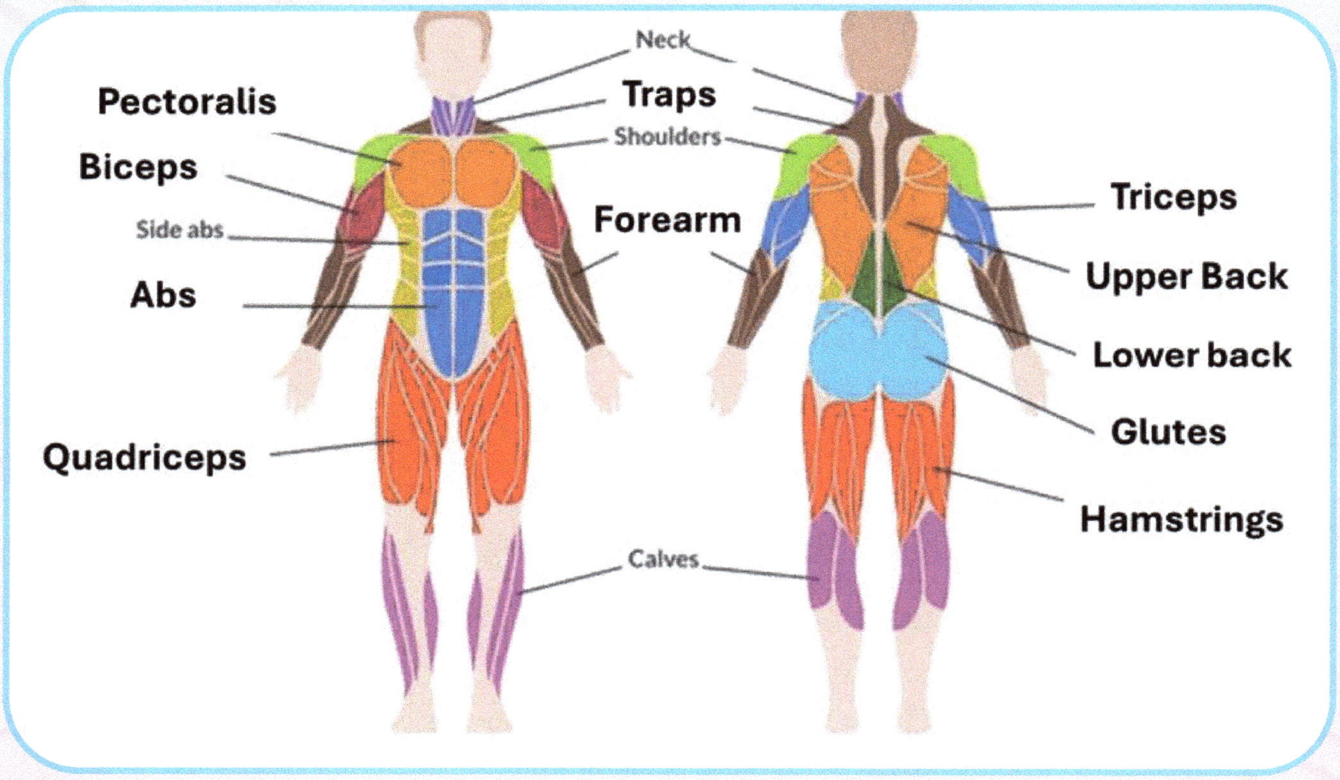

WHAT UPPER BODY MUSCLES DO

Deltoids: Shoulder muscles that help lift and rotate the arms.

Biceps: Front upper arm muscles that help bend the elbow.

Triceps: Back upper arm muscles that help straighten the elbow.

Pectorals: Chest muscles that help move the shoulder joint.

WHAT LOWER BODY MUSCLES DO

Quadriceps: Front thigh muscles that help straighten the knee and flex the hip.

Hamstrings: Back thigh muscles that help bend the knee and extend the hip.

Glutes: Buttock muscles that help with hip movement and stability.

Calves: Lower leg muscles that help with ankle movement and balance.

INTERESTING MUSCLES AND THEIR SCIENTIFIC NAMES

Now that we know the basic names and a few of the modern terms for muscles, let's review some of the scientific names and a few fun facts about muscles.

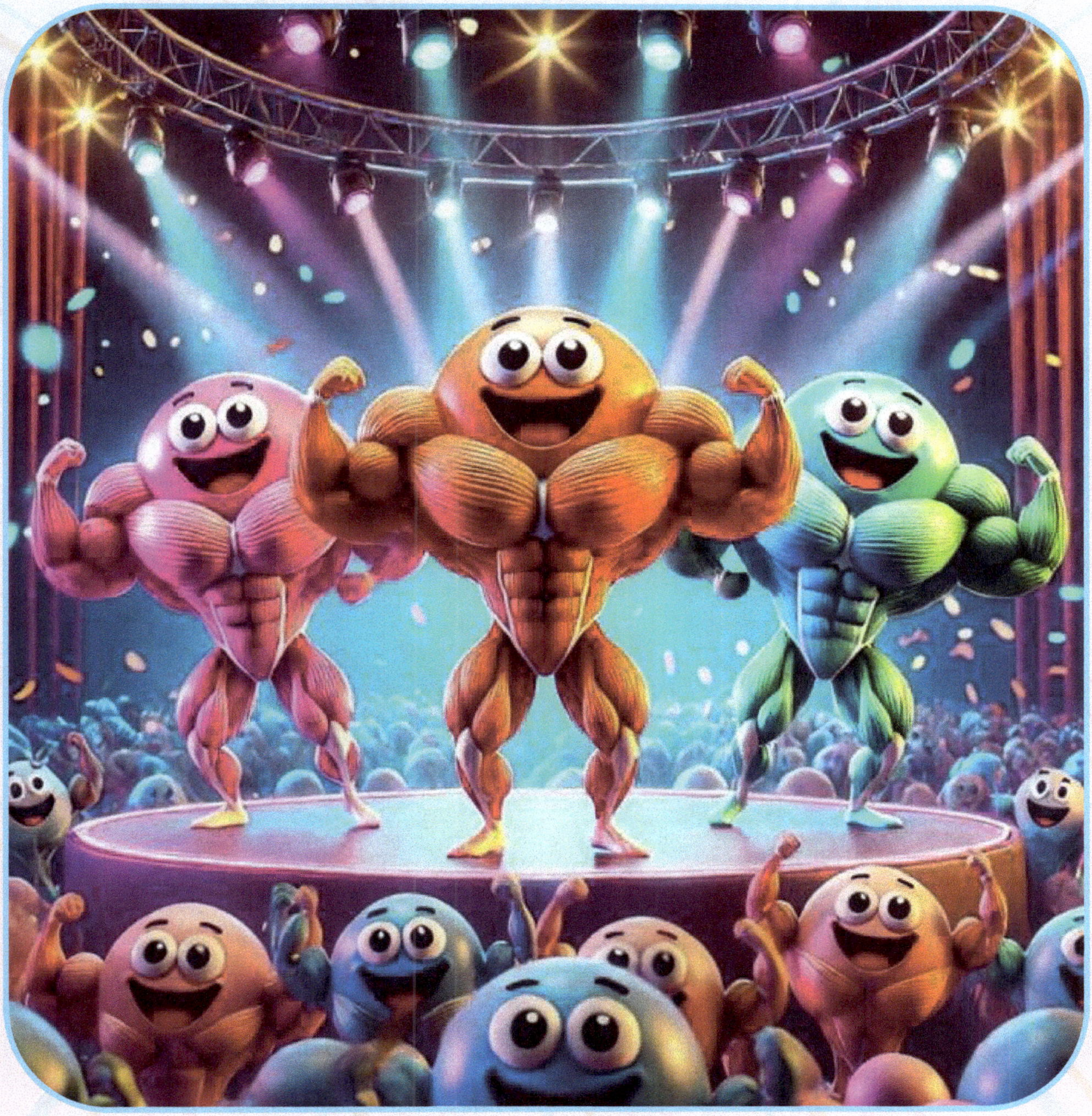

TRAPEZIUS

The trapezius muscle, commonly referred to as "traps," is the large, muscle that extends across the back of the neck and shoulders. It's involved in a variety of movements and plays a significant role in maintaining posture and allowing for a wide range of shoulder and neck motions.

Shape: The trapezius muscle is shaped like a triangle and resembles a giant starfish.

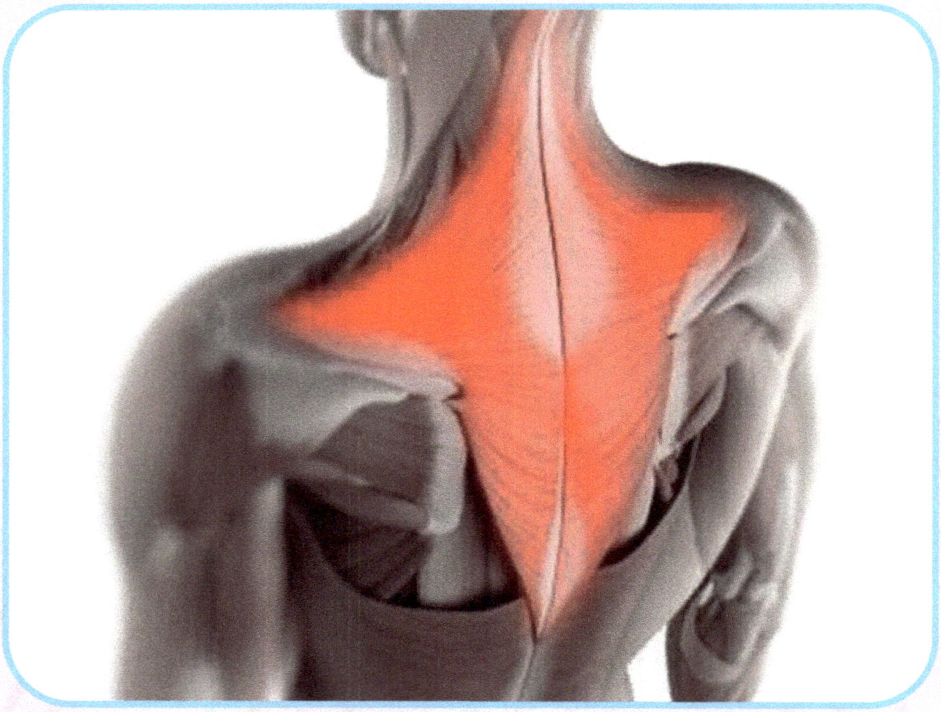

LATISSIMUS DORSI

The **latissimus dorsi**, often referred to as the "**lats,**" is a large, flat muscle located on the back. It's interesting because it is one of the widest muscles in the human body. This wide muscle plays a key role in various movements involving the upper body and the arms.

Shape: It is shaped like an upside-down arrowhead and resembles bat wings.

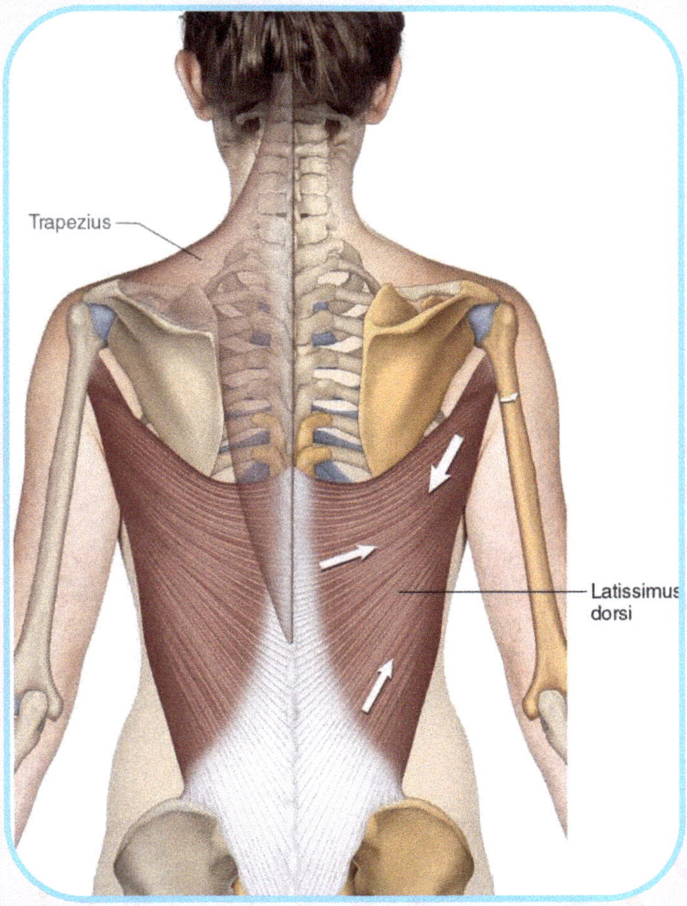

GLUTEUS MAXIMUS

The **gluteus maximus** (bottom) is the largest muscle in the body. It is responsible for the movement of the hip and thigh, including activities like walking, running, and climbing stairs.

Shape: The glutes have an onion shape to them.

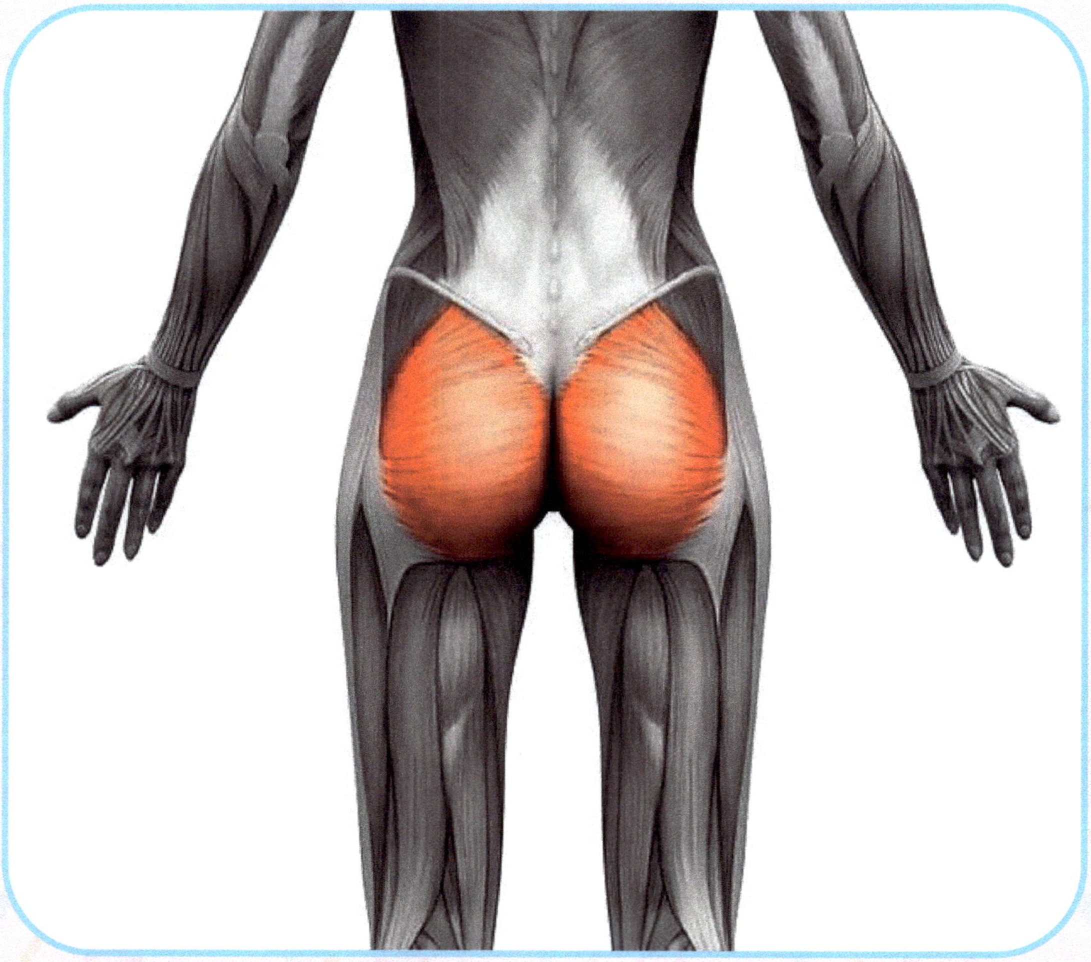

THE DIAPHRAGM

The **diaphragm** is a large, dome-shaped muscle located at the base of the lungs, separating the thoracic cavity (which contains the heart and lungs) from the abdominal cavity. It plays a crucial role in the respiratory system and is the primary muscle responsible for breathing.

Shape: It has a dome shape and looks like a mushroom.

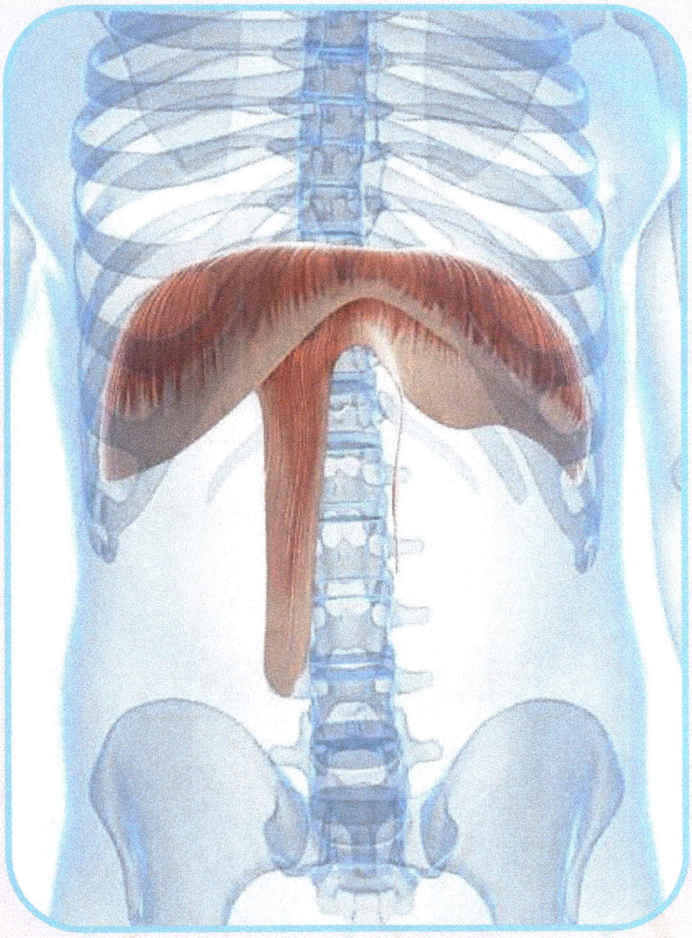

RHOMBOID MUSCLES

The **rhomboid muscles** are a pair of muscles located in the upper back, beneath the trapezius. These muscles are important for the movement and stabilization of the shoulder blades (scapulae).

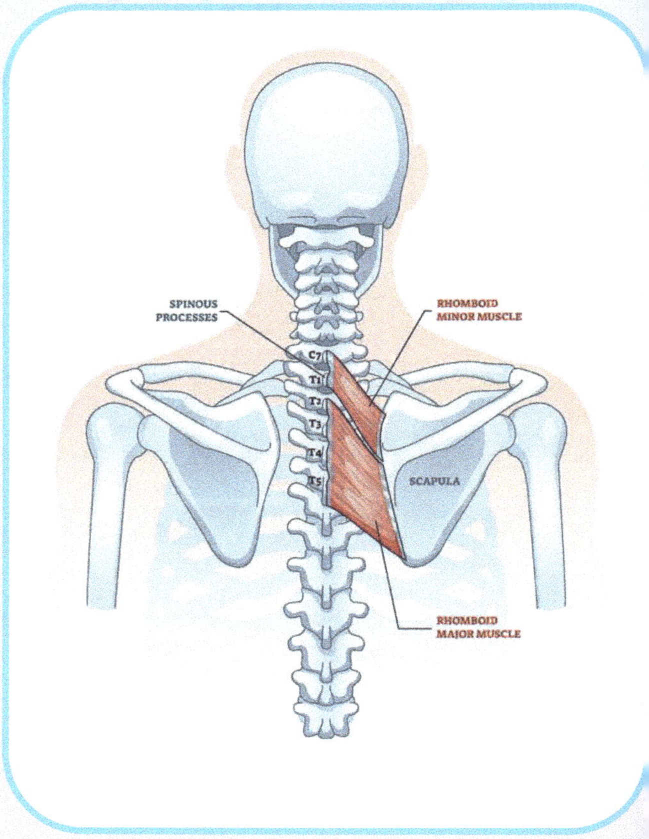

Shape: Both muscles are roughly diamond-shaped, but nothing cool to compare them with. What shape do you think they look like?

WHAT PART OF THE BODY HAS THE MOST MUSCLES?

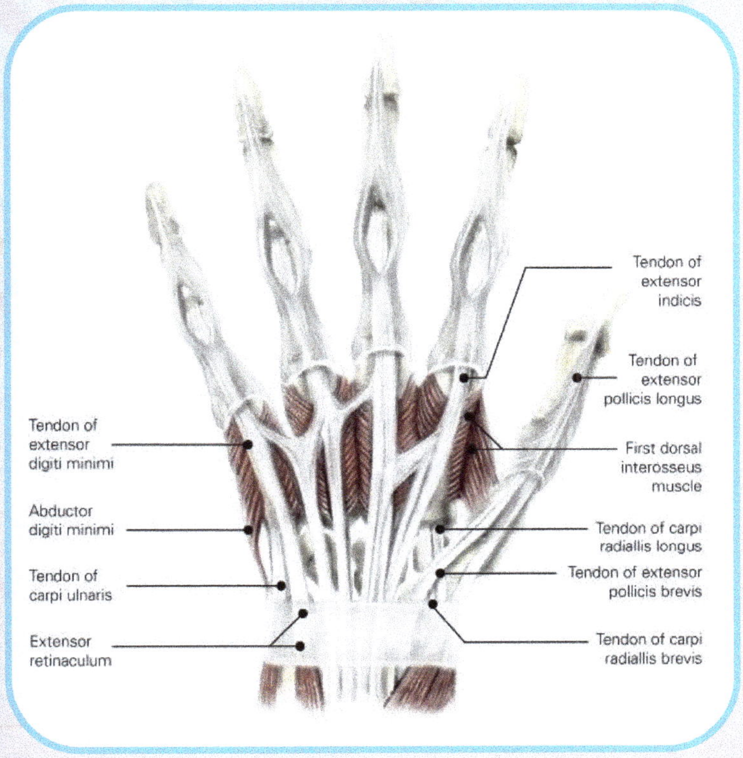

The area of the body with the most muscles is the hands. Each hand contains about 34 muscles that work together to perform complex and precise movements. These muscles are responsible for the movement and fine motor skills required for tasks like typing, playing musical instruments, and grasping objects.

Muscles in the Hand

- **Extrinsic Muscles:** These muscles originate in the forearm and extend into the hand. They control the main movements of the fingers and wrist.
- **Intrinsic Muscles:** These are smaller muscles located entirely within the hand. They allow for fine motor skills and more precise movements, such as gripping and manipulating small objects.

FACE MUSCLES

The facial muscles can convey many emotions and feelings without us speaking a single word. They help us speak on a universal level in a way that anyone can understand no matter what country you are in. This section is dedicated to the face.

FACE MUSCLES

The face contains many muscles that are essential for facial expressions, chewing, speaking, and other functions. Let's review a few of these facial muscles:

The names of face muscles like many other muscles come from the old Latin and Greek languages. They can include two or three names and involve up to five syllables.

Many of these Latin and Greek names can be challenging to pronounce but fun to hear and learn.

Not many people use these terms, but they can be impressive, fun, and exciting to do so. See if you can pronounce them.

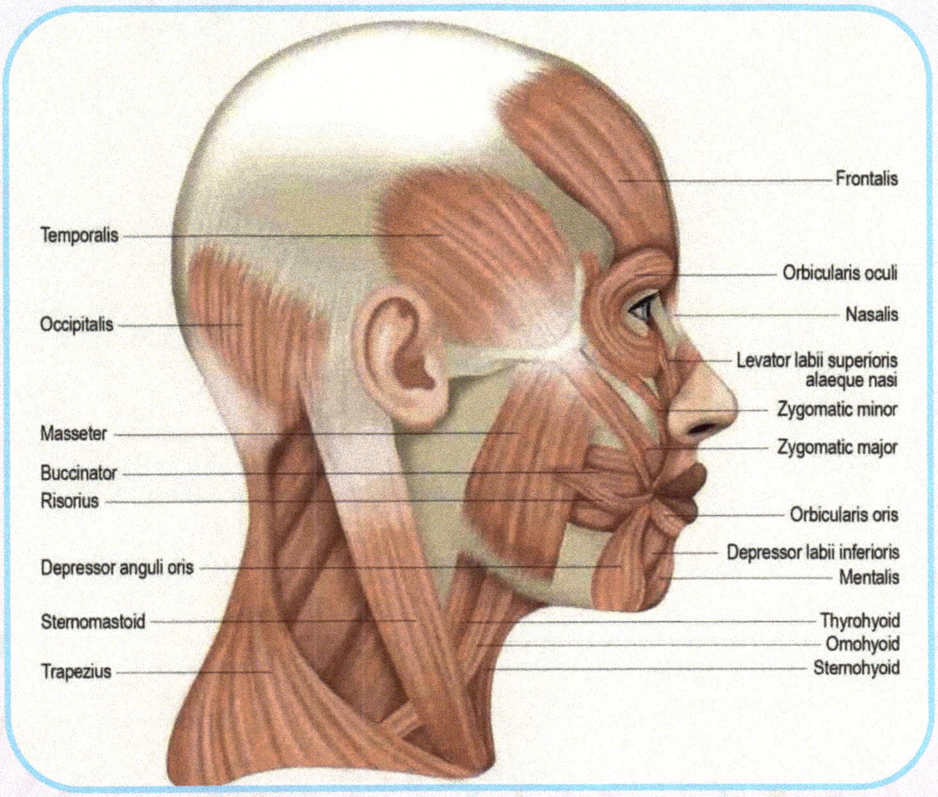

FRONTALIS MUSCLE

Frontalis Muscle: This muscle raises the eyebrows and wrinkles the forehead.

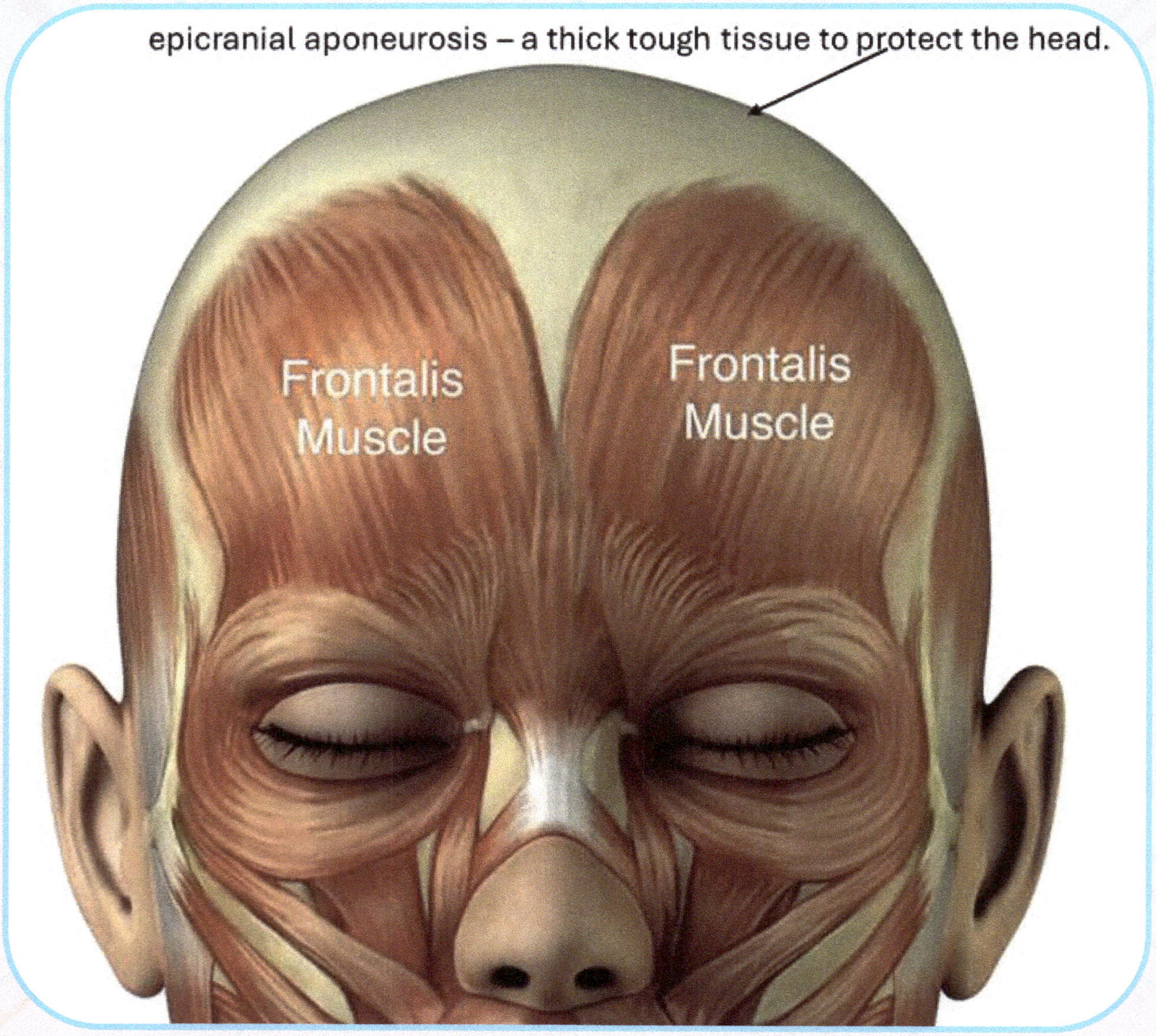

Fun Fact

> "The face is home to more than 40 muscles, each responsible for a wide array of expressions and movements."

ORBICULARIS OCULI

Orbicularis Oculi closes the eyelids. These muscles can help with the widening or narrowing of the eyes. They have a circular shape to them.

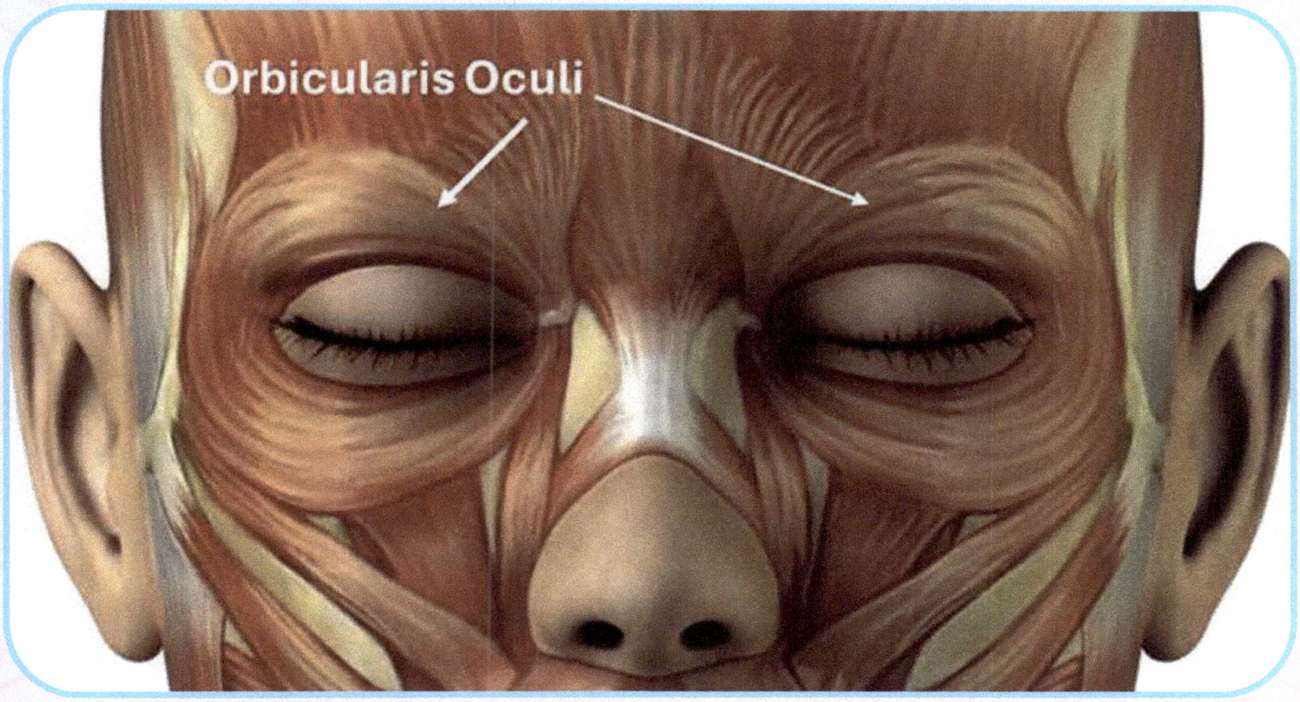

Fun Fact

"Your eye muscles are the fastest muscles in your body. They allow you to move your eyes in different directions in the blink of an eye—literally!"

ZYGOMATICUS MAJOR AND MINOR

Zygomaticus Major and Minor: These muscles raise the corners of the mouth, which can help in creating a beautiful smile or a frown.

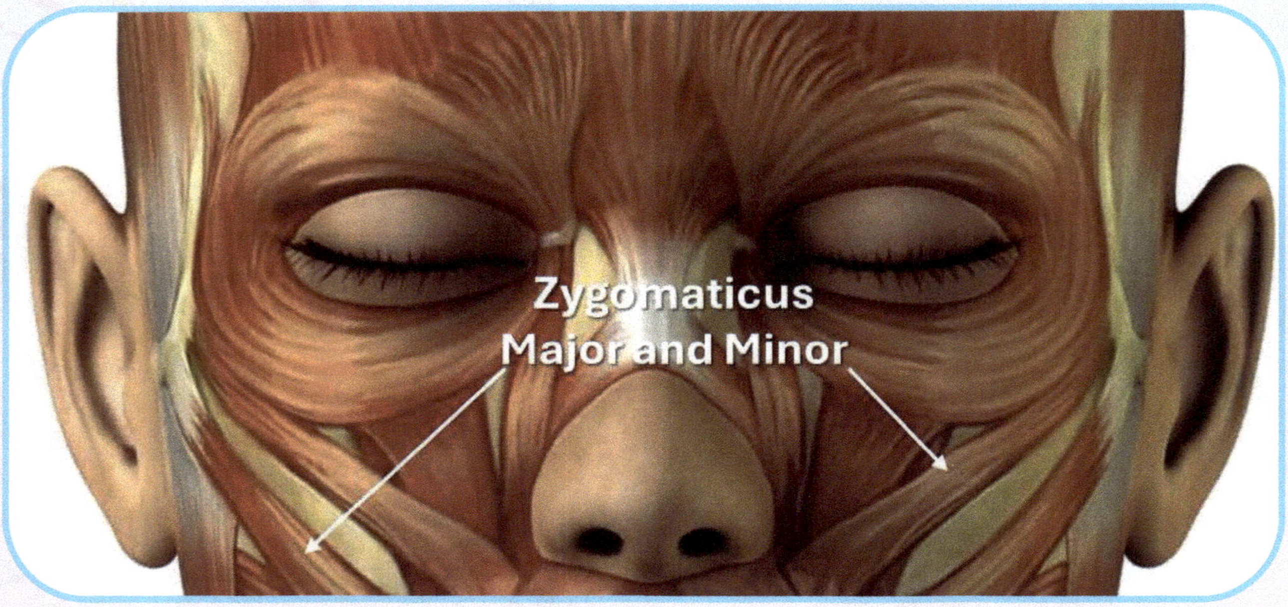

LEVATOR LABII SUPERIORIS

Levator Labii Superioris: This muscle raises the upper lip.

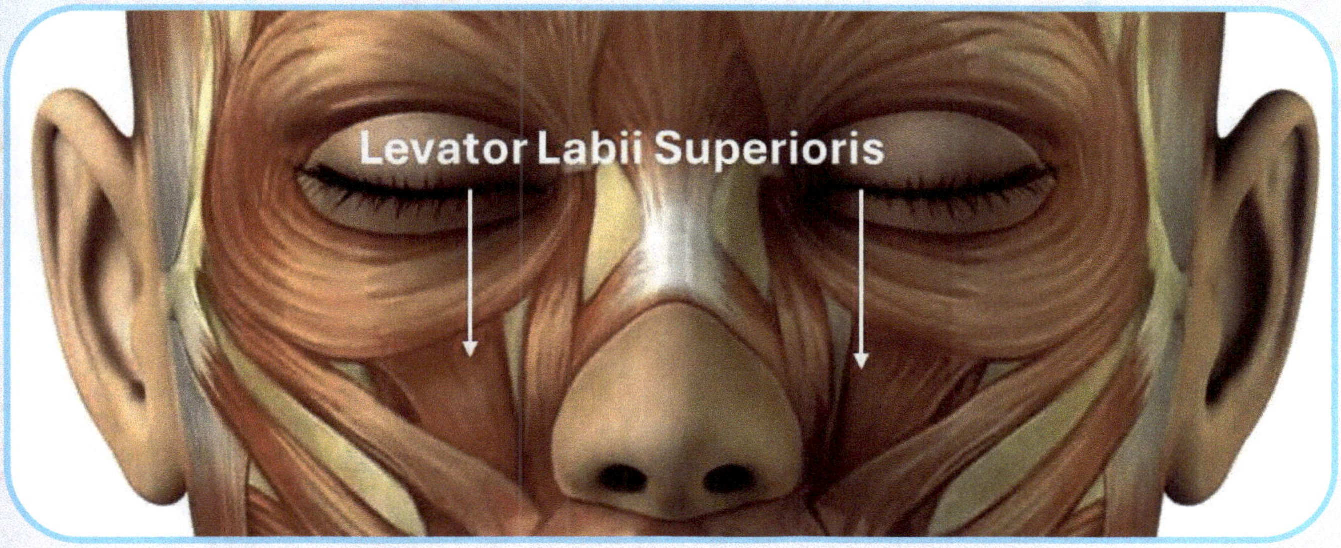

MASSETER MUSCLE

The **masseter muscle** is one of the main muscles involved in the process of chewing (mastication). It's a powerful and important muscle located in the jaw and supports a variety of mouth movements.

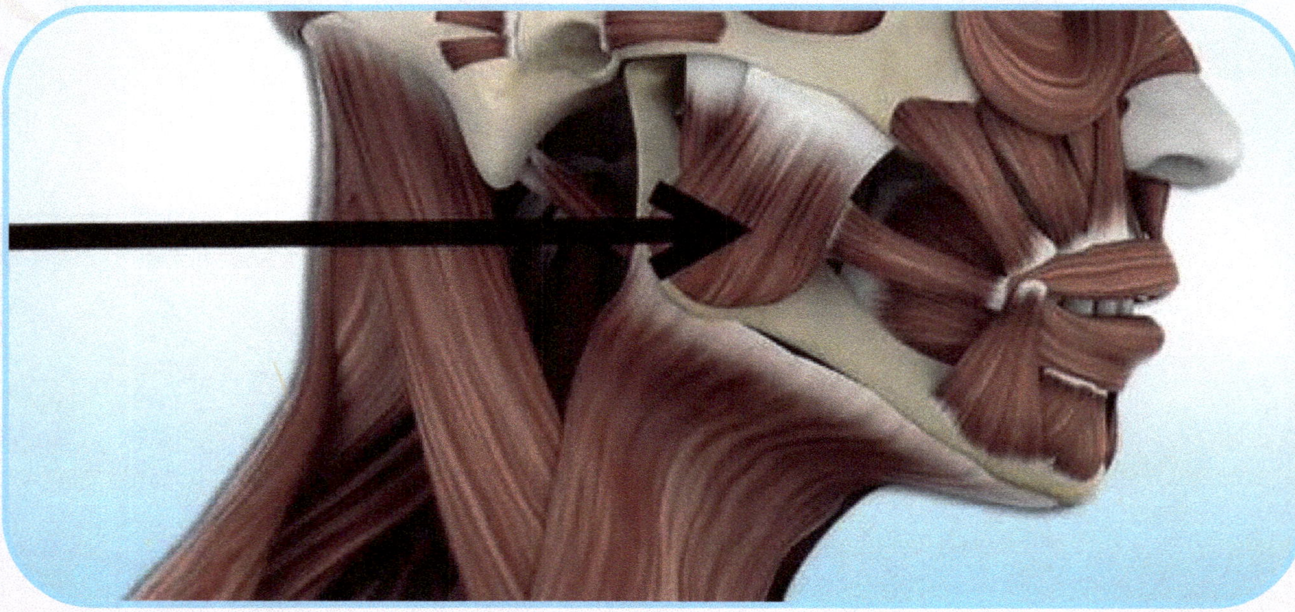

FACE MUSCLES FUN FACTS

Expressive Power: The human face has over 40 muscles, allowing us to make more than 7,000 distinct facial expressions. These muscles work together to convey a wide range of emotions, from joy and surprise to sadness and anger.

Universal Language: Smiling is a universal sign of happiness and friendliness recognized across all cultures. Even babies born blind smile in response to the sound of a human voice, indicating that smiling is an innate behavior.

Health Benefits: Smiling releases endorphins, serotonin, and natural painkillers, which can help reduce stress and improve your mood. It's often said that a smile is the best medicine!

WALKING MUSCLES

Walking is a muscle activity that we learn to do very early in our lives. It involves a coordinated effort with many muscles throughout the body. Let's review the way we walk, and the primary muscles involved.

THE GAIT

The scientific term for walking is called gait. It is a process that involves the leg **stance** and the leg **swing**. This activity requires around 7 movements that happen without us even thinking much about it

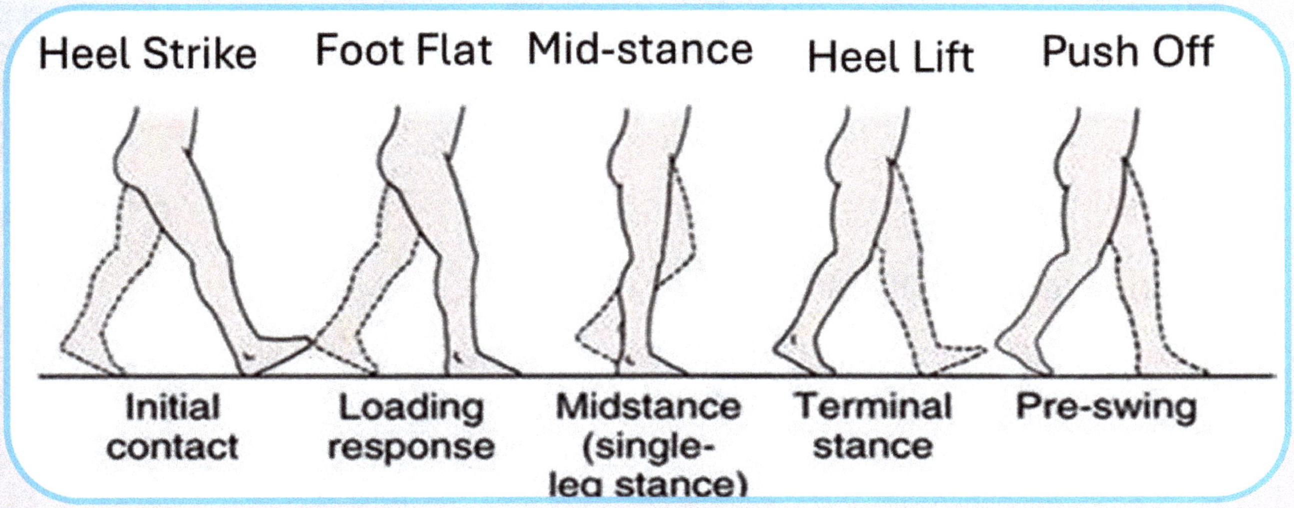

GAIT – THE LEG STANCE

Stance Phase (60% of the gait cycle):

1. **Heel Strike:** The heel contacts the ground, and the leg begins to absorb the body's weight.
2. **Loading Response:** The foot flattens, and the body shifts weight onto the leg.
3. **Midstance:** The body weight is directly over the supporting leg.
4. **Terminal Stance:** The heel lifts off the ground as the body weight moves forward.
5. **Pre-swing:** The toes push off, preparing for the swing phase.

Did you Know?

"Walking involves more than 200 muscles all working together in a coordinated manner."

GAIT – THE LEG SWING

The swinging of the legs is also an important part of walking. Let's review how the legs swing.

Swing Phase (40% of the gait cycle):

1. **Initial Swing:** The leg lifts off the ground and moves forward.

2. **Mid-swing:** The leg swings through beneath the body.

3. **Terminal Swing:** The leg extends, preparing for the next heel strike.

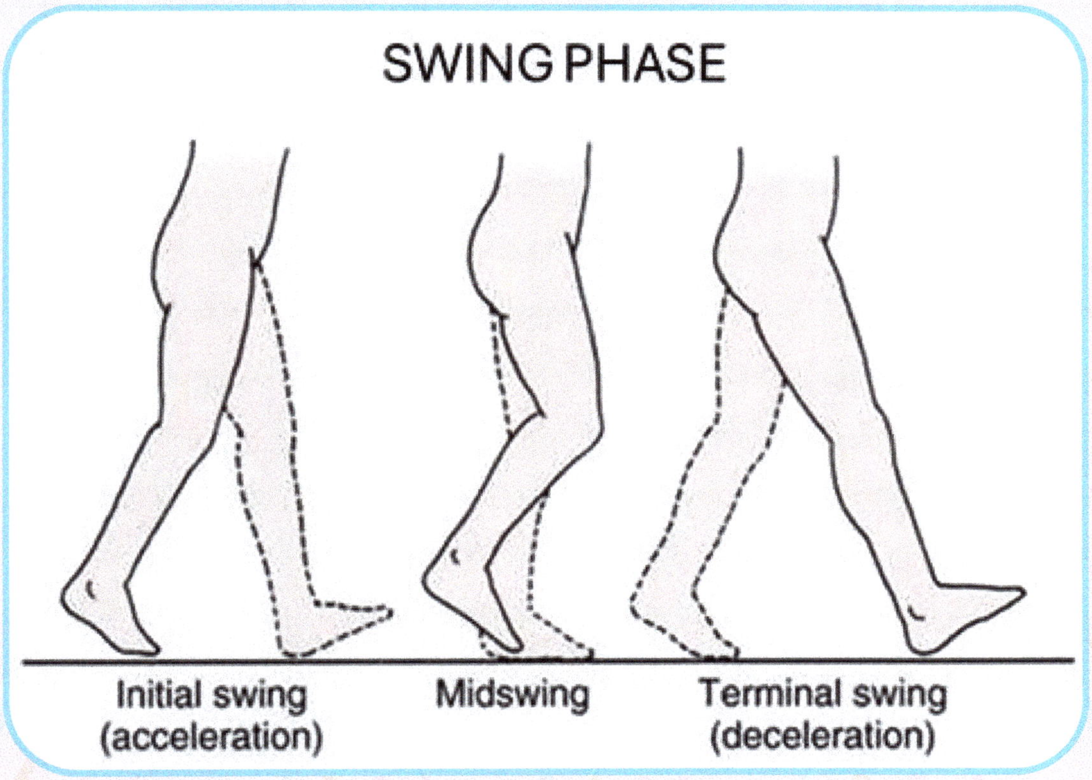

WALKING KEY MUSCLES

These are the key muscles that are involved in walking.

Quadriceps: Extends the knee during the stance phase.

Hamstrings: Flexes the knee during the swing phase.

Glutes: Extend and stabilize the hip during the stance phase.

Calves: Push off the ground during the pre-swing phase.

Tibialis Anterior: Flexes the foot during the swing phase to clear the ground.

Obliques: The oblique muscles are part of your abdominal muscles, located on the sides of your abdomen. They help with side bending and twisting your torso

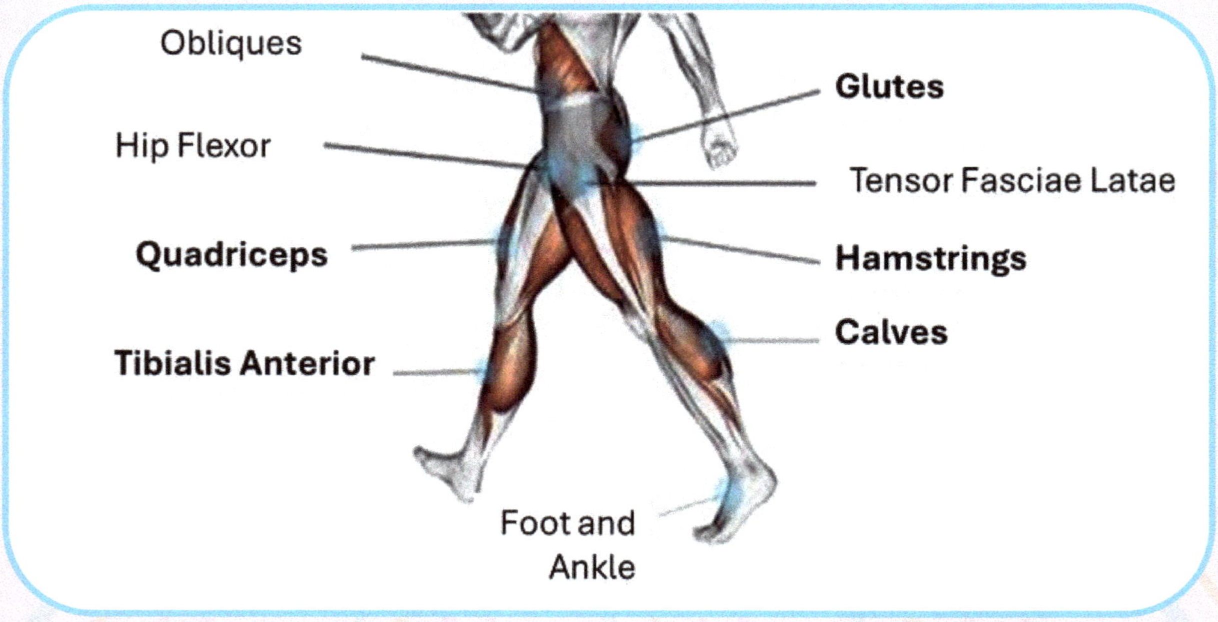

LEG MUSCLES

In total, there are over 50 muscles in the leg, each playing a vital role in movement, stability, and balance. If any one of these muscles gets injured, it can impact your ability to move properly.

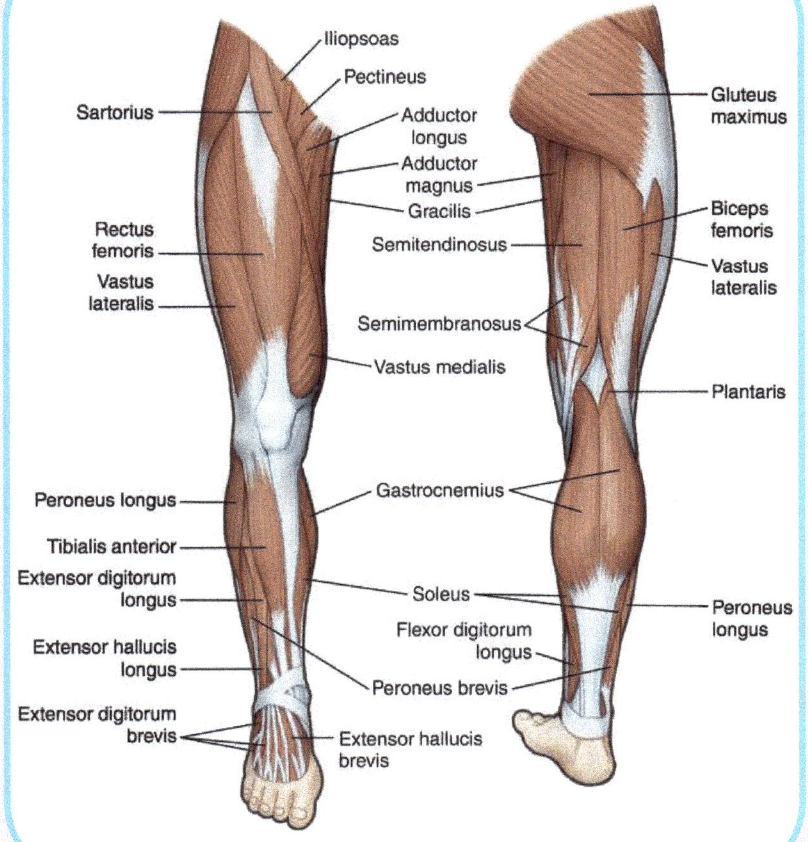

Did you Know?

Longest Muscle: The sartorius, running from the hip to the inner knee, is the longest muscle in the human body. It's involved in movements like crossing your legs.

LEG MUSCLE FUN FACTS

1. Strongest Muscle:

The gluteus maximus, which forms the bulk of your buttocks, is one of the strongest muscles in the body. It's essential for movements like running, climbing stairs, and standing up from a seated position.

2. Most Endurance:

The soleus, a muscle in your calf, is known for its endurance. It's highly efficient at working over long periods without fatigue, making it crucial for activities like walking and standing.

3. Largest Muscle Group:

The quadriceps, or "quads," are the largest muscle group in your body. Comprising four muscles (rectus femoris, vastus lateralis, vastus medialis, and vastus intermedius), they are vital for extending the knee and powerful leg movements.

CHEWING MUSCLES

Chewing is another important muscle activity we perform every day. We learn to chew very early in our lives. But chewing requires a team of muscles to work together at the same time.

THE CHEWING PROCESS

1. **Preparation:** Food is placed in the mouth, and the lips are sealed to keep it inside.
2. **Initial Breakdown:** The teeth, primarily the molars (rear grinders), break down the food into smaller pieces through crushing and grinding motions.
3. **Saliva Production:** Salivary glands secrete saliva, which moistens the food, making it easier to chew and swallow. Saliva also contains enzymes that begin the chemical digestion of carbohydrates.
4. **Mastication:** The muscles of mastication (chewing) work together to move the jaw up and down, and side-to-side, ensuring that food is thoroughly mixed with saliva and broken down into smaller pieces.
5. **Bolus Formation:** The tongue helps form the chewed food into a cohesive mass called a bolus, which is then ready to be swallowed.

MUSCLES INVOLVED IN CHEWING

1. **Masseter:** One of the sturdiest muscles in the body, responsible for elevating the jaw and crushing food.
2. **Temporalis:** Assists in closing the jaw and grinding food.
3. **Lateral Pterygoid:** Also helps in opening the mouth and moving the jaw from side to side.
4. **Buccinator:** Compresses the cheeks, helping to position food between the teeth.

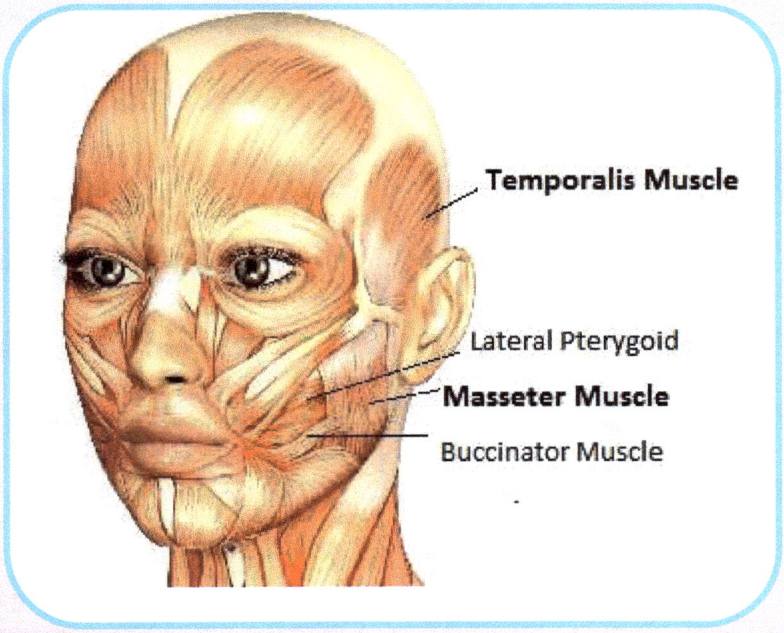

Did you Know?

"The masseter muscle is considered one of the strongest muscles in the human body relative to its size. It can produce a bite force of up to 200 pounds on the molars."

IS THE TONGUE A MUSCLE?

Answer: Yes, the tongue is indeed a muscle, but it's not just any muscle—it's a fascinating and highly versatile one. The tongue is composed of a group of muscles known as the intrinsic and extrinsic muscles.

1. **Intrinsic muscles:** These muscles are located entirely within the tongue and are responsible for changing its shape, allowing you to perform actions like curling, flattening, or pointing the tongue.
2. **Extrinsic muscles:** These muscles originate outside the tongue but are inserted into it. They help move the tongue in different directions, such as sticking it out, moving it side to side, and aiding in speech and swallowing.

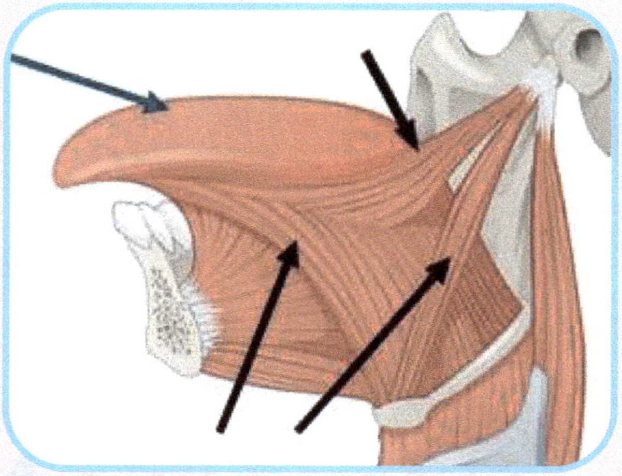

CHEWING FUN FACTS

Symmetry and Asymmetry: While our facial muscles work symmetrically (equally), we tend to favor one side while chewing, leading to slight asymmetry (not equal). This preference can be influenced by factors like hand dominance and dental health.

Grinding Teeth: Some people grind their teeth at night (a condition known as bruxism), which heavily engages the chewing muscles. This can lead to jaw pain and other issues.

Facial Expressions: Chewing muscles also play a role in facial expressions. For instance, the masseter and temporalis muscles can contribute to expressions of tension or anger.

ARM MUSCLES

The arms play a significant role in our daily activities. Without them, our lives would be very different. This section will review some of the major muscles of the arm.

ARM POSITIONS

In anatomy, the position of the arms is important when describing muscles. The anterior is when the palms are facing upwards. This is known as supination.

The posterior is when the palms are facing down. This is known as pronation.

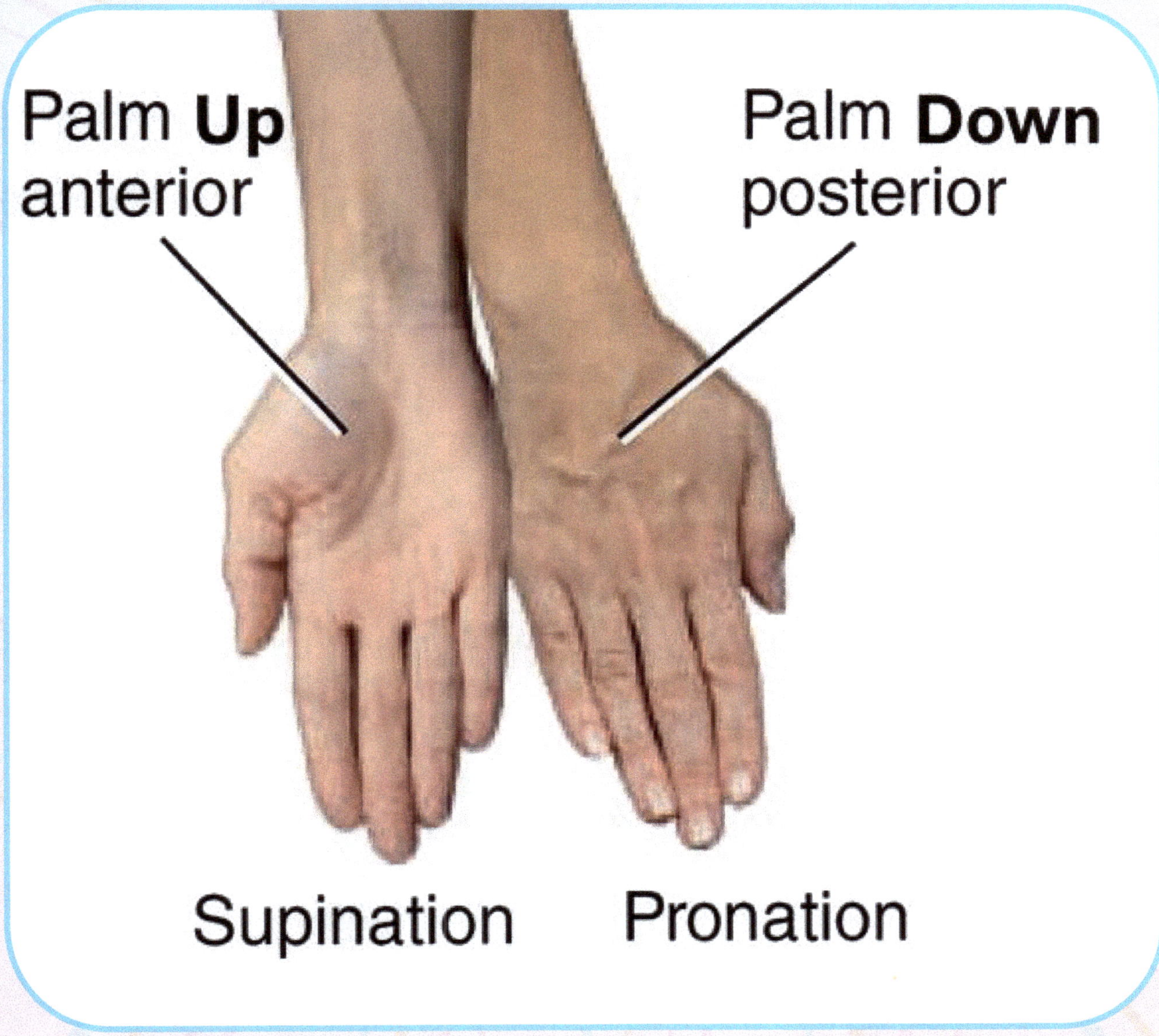

TYPES OF ARM MUSCLES

In total, there are about 20 primary muscles in each arm. However, there are additional smaller muscles and muscle groups that play supportive roles in the function and movement of the arm.

Arm muscles can be divided into two basic categories: **Flexors** and **Extensors.**

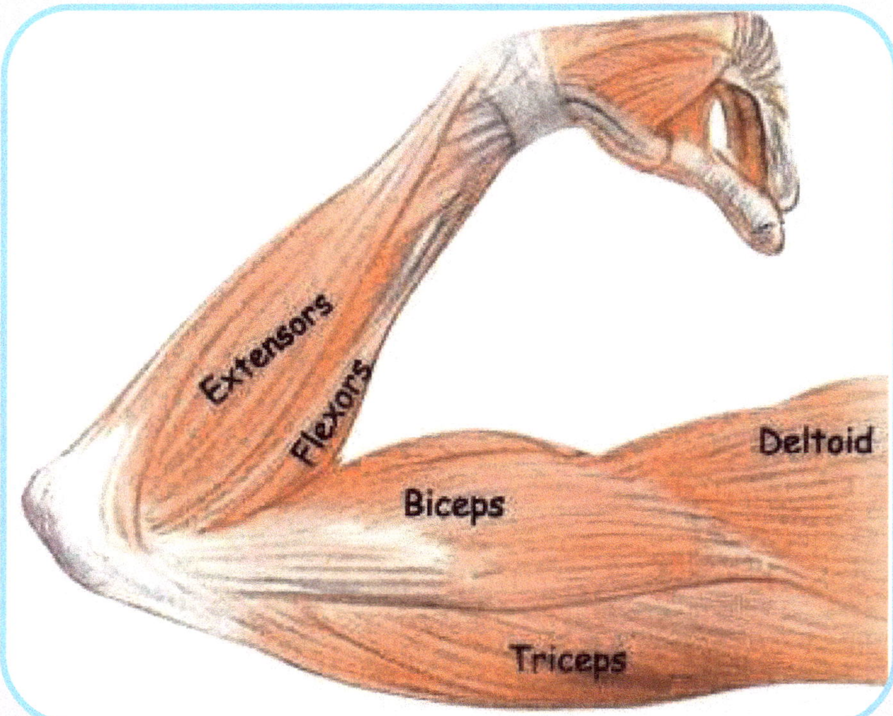

Did you Know?

"Muscles have the incredible ability to "remember" movements through repeated practice. This phenomenon is known as muscle memory and the reason why athletes and musicians can perform complex tasks with ease after extensive training."

FLEXOR ARM MUSCLES

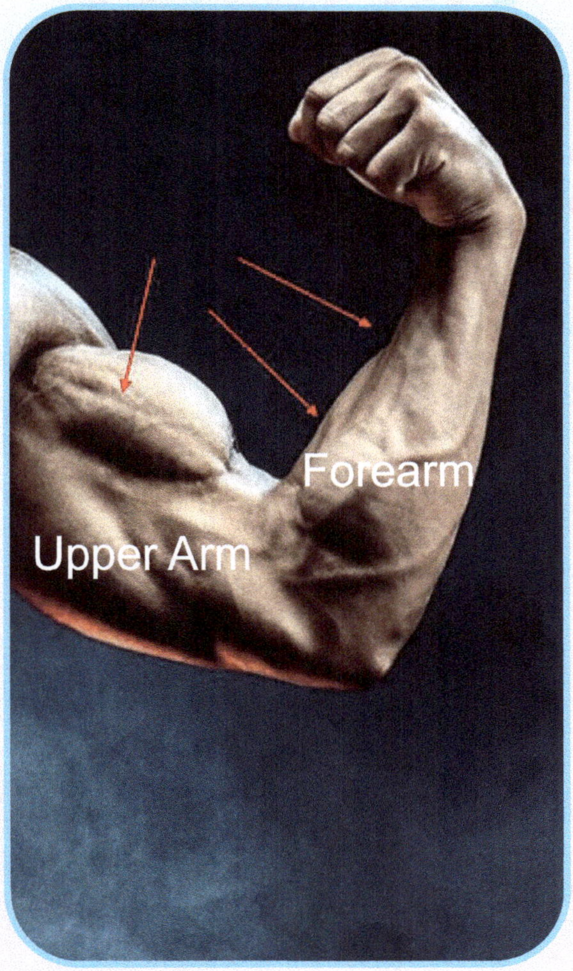

Flexor arm muscles are responsible for bending or flexing the joints in the arm, particularly at the elbow and wrist.
These muscles are primarily located on the front (anterior) side of the forearm and upper arm.

UPPER ARM FLEXORS

There are two major flexor muscles in the upper arm.

1. **Biceps Brachii:** This muscle has two heads (short and long) and is responsible for flexing the elbow and supinating the forearm (rotating the palm upwards).
2. **Brachialis:** This muscle is located underneath the biceps. It helps in flexing the elbow.

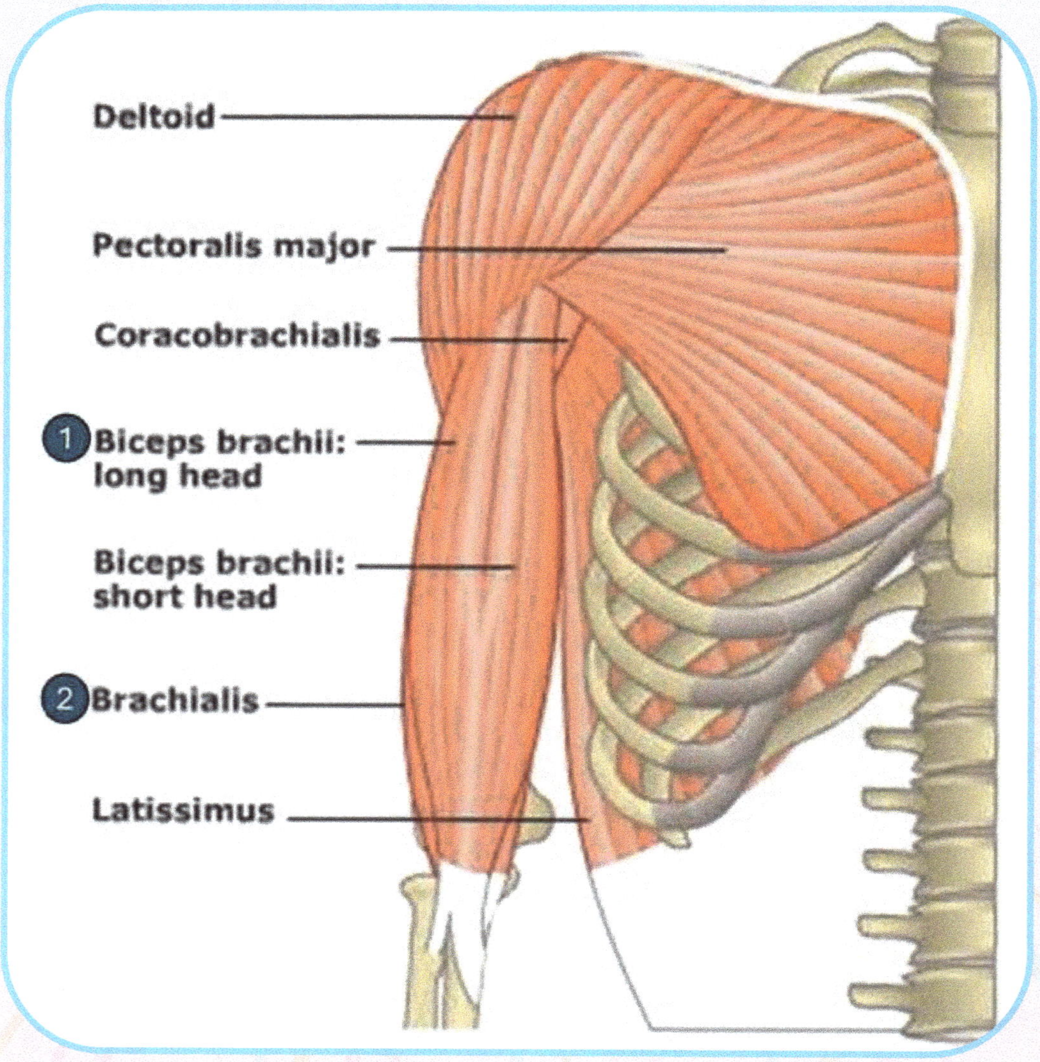

FOREARM FLEXORS

The forearm has many more flexor muscles than the upper arm. Here are a few:

1. **Flexor Carpi Radialis:** Helps in flexing and abducting the wrist.
2. **Flexor Carpi Ulnaris:** Assists in flexing and adducting the wrist.
3. **Palmaris Longus:** Flexes the wrist.
4. **Flexor Digitorum Superficialis:** Flexes the fingers.
5. **Pronator Teres:** Assists in pronating the forearm (rotating the palm downwards).

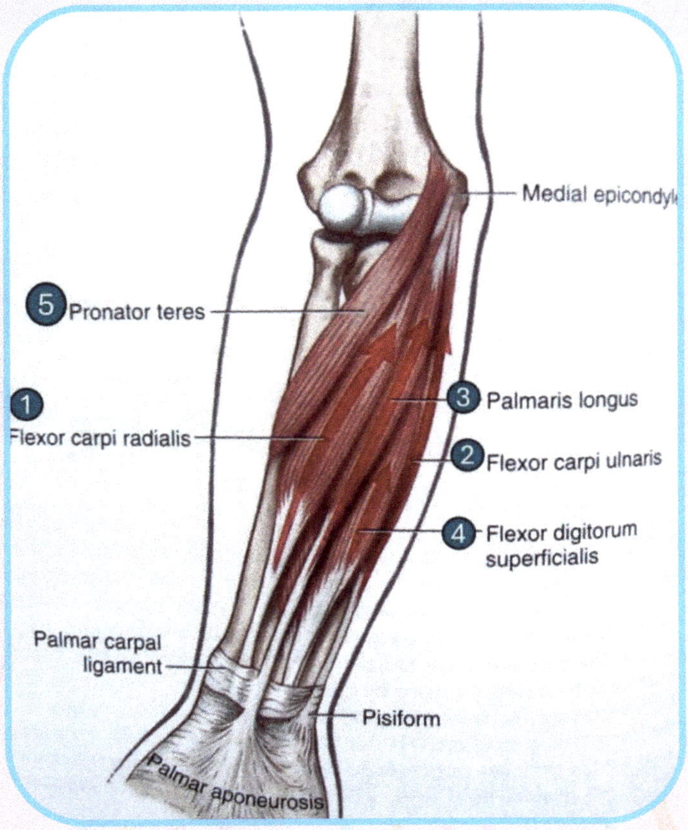

EXTENSOR MUSCLES

Extensor muscles in the arm are responsible for extending or straightening the joints, particularly at the elbow and wrist. These muscles are primarily located on the back (posterior) side of the forearm. Here are the key extensor muscles in the arm:

UPPER ARM EXTENSOR MUSCLES

While there are fewer extensor muscles in the upper arm compared to the forearm, the primary muscle responsible for extension at the elbow is:

Triceps Brachii: This muscle has three heads and is responsible for extending the elbow.

FOREARM EXTENSOR MUSCLES

Most of the extensor muscles are in the forearm. Here are a few:

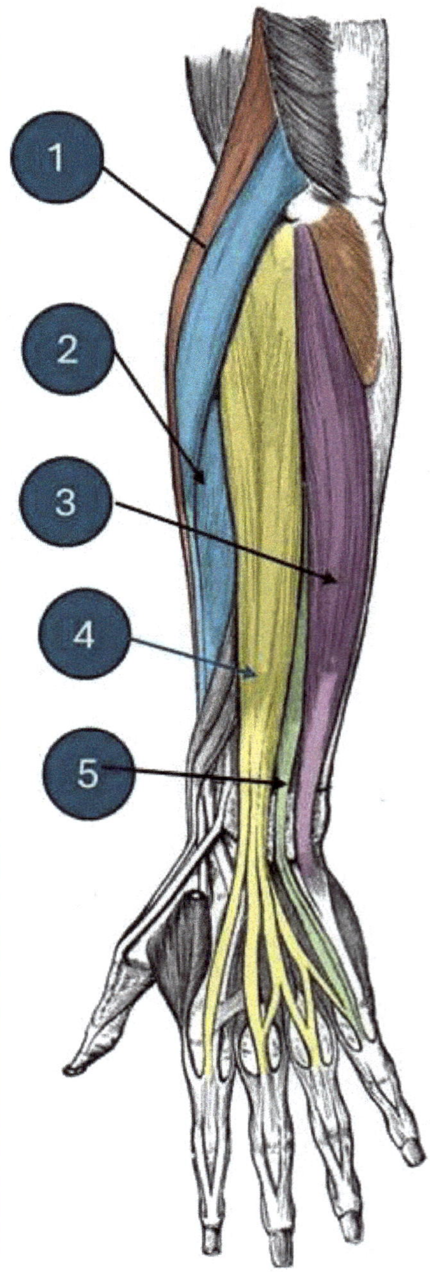

1. **Extensor Carpi Radialis Longus:** Extends and moves the wrist upwards.

2. **Extensor Carpi Radialis Brevis:** Extends and moves the wrist upwards.

3. **Extensor Carpi Ulnaris:** Extends and helps move the wrist downwards.

4. **Extensor Digitorum:** Extends the fingers.

5. **Extensor Digiti Minimi:** Specifically extends the little finger.

ALL MUSCLE NAMES

Learning all the names of muscles can be challenging yet fun. Study the diagram to learn more about the various names of muscles.

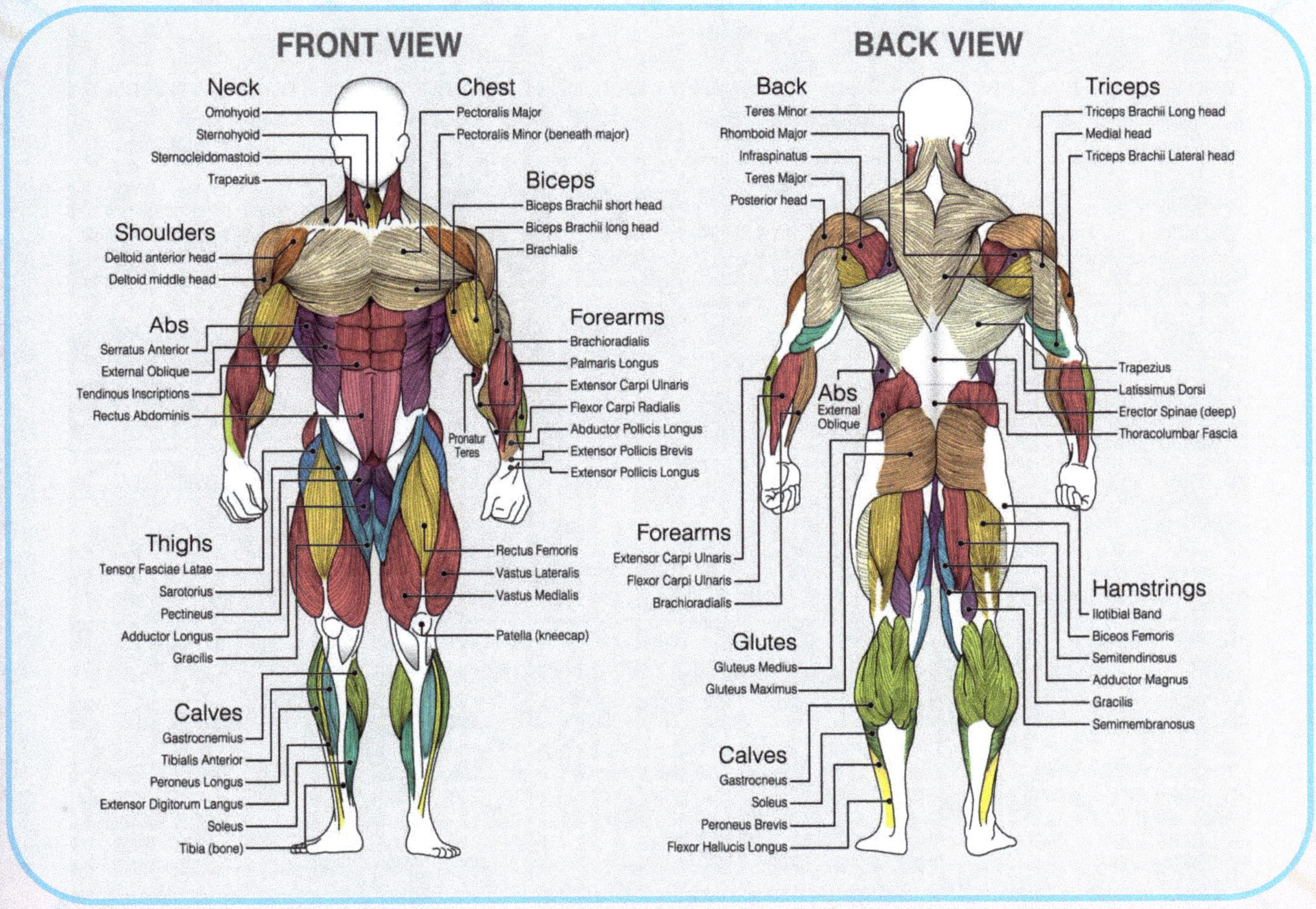

KNOWLEDGE CHECK #1

Practice is the key to memorization. Try to fill in the blanks with the correct muscle name.

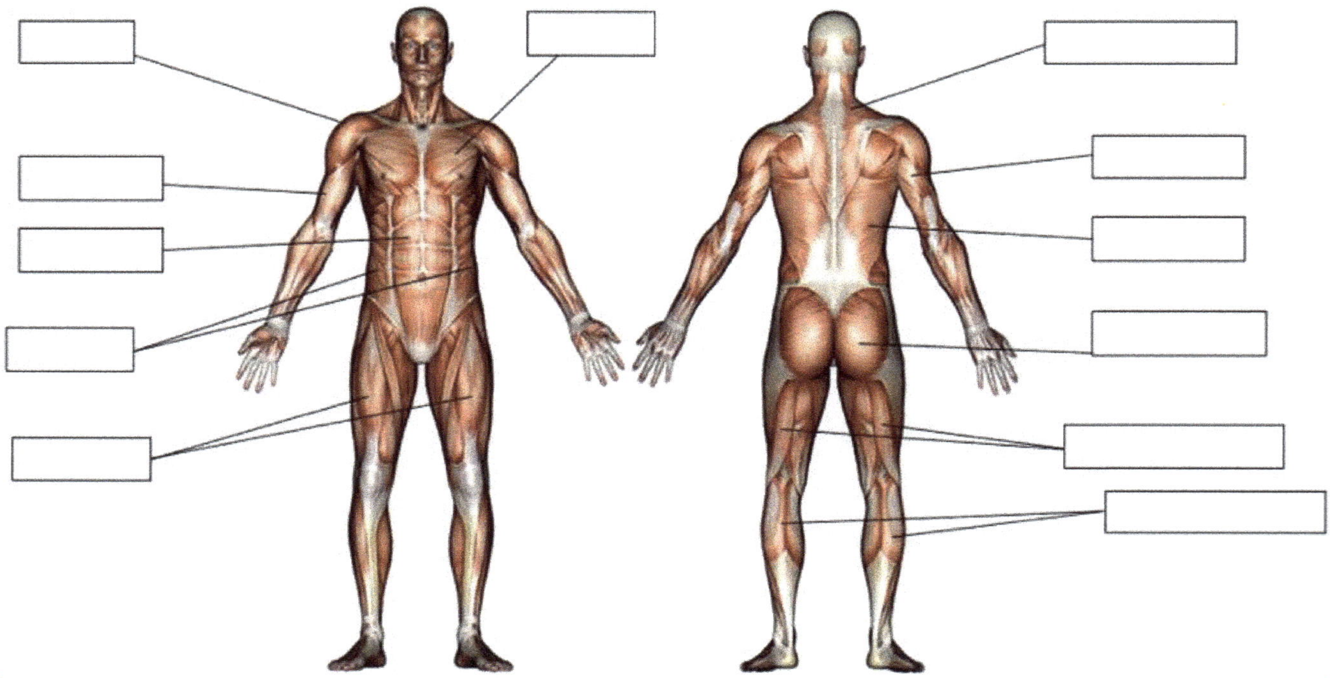

KNOWLEDGE CHECK #2

Practice and repetition can help you to remember the anatomical names of muscles. This one has more muscle groups. See if you can name the muscles on this diagram.

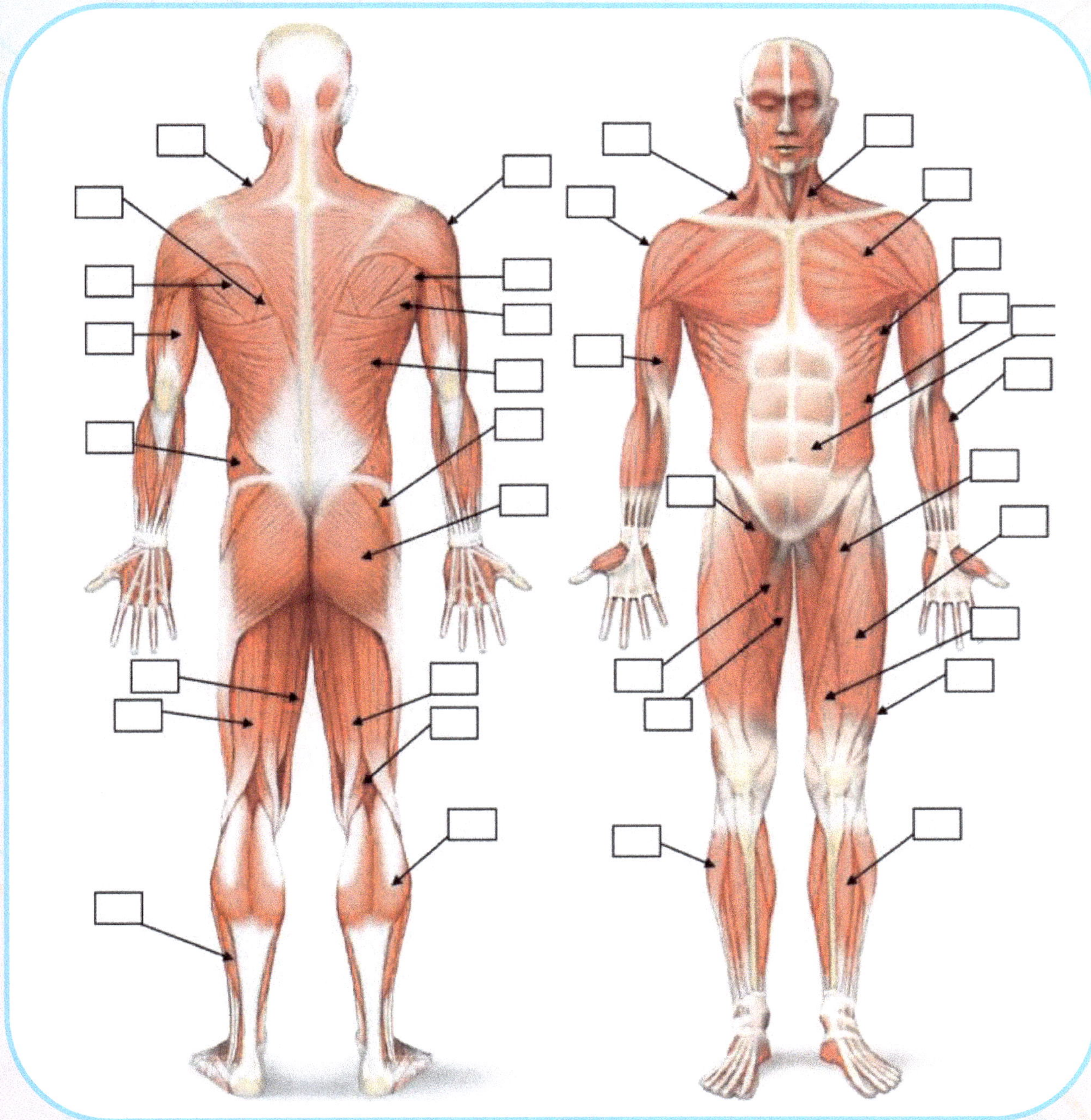

CONCLUSION

I hope you have enjoyed reading this book. We learned many interesting facts about muscles, what they do, how they function, what they are made of, and their basic and scientific names. Do not worry about trying to learn all the names of the muscles all at once. As you practice and continue your studies, the names will become easier for you to remember.

Stay safe and never stop learning!

Other STEM Books by ZB Tucker

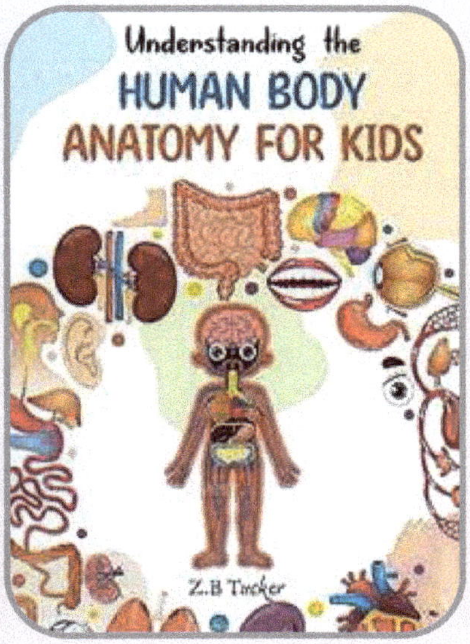

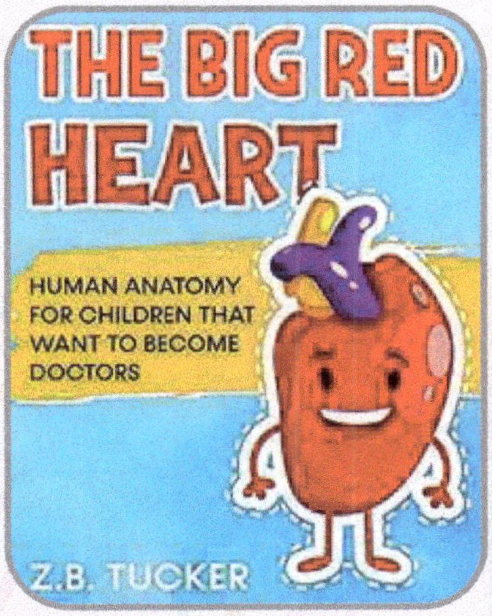

Author's Bio

The author is an educator and a passionate lover of children. He recieved his Bachelor of Arts in Biology from Edgewood College in Madison. Wisconsin and a secon B.A. in Healthcare Management from Southern Illinois University Carbondale where he also completed his M.A. in Engineering. He has a wealth of valubale experience dealing with children of varying age ranges. The author does not only love children, but he understands that when they have the best form of education, it has a far-reaching effect on the wider society. As a result, he is on a mission to make sure that children learn as much as they can.

References

Gray's Anatomy for Students. Richard L. Drake. March 24, 2023. Elsevier, 5th edition

Dorland's Illustrated Medical Dictionary. Dorland. Dec 30, 2019. Elsevier (33rd ed.)

Fun Facts About the Muscular System You Didn't Know. Healthline. James Roland. Updated May 1, 2024. https://www.healthline.com/health/fun-facts-about-the-muscular-system

Human Muscle Systems. Britannica. https://www.britannica.com/science/human-muscle-system

Interior images by Shutterstock, Microsoft Stock Images, iStock, and Zach's Creative Space

www.ingramcontent.com/pod-product-compliance
Lightning Source LLC
LaVergne TN
LVHW070857210926
848482LV00016B/962